LOVE WINS

LOVE WINS

The Lovers and Lawyers Who Fought the Landmark Case for Marriage Equality

DEBBIE CENZIPER AND
JIM OBERGEFELL

WILLIAM MORROW
An Imprint of HarperCollins*Publishers*

HarperCollins books may be purchased for educational, business, or sales promotional use. For information please e-mail the Special Markets Department at SPsales@harpercollins.com.

FIRST EDITION

Designed by Bonni Leon-Berman

Library of Congress Cataloging-in-Publication Data has been applied for.

ISBN 978-0-06-245608-3

16 17 18 19 20 OV/RRD 10 9 8 7 6 5 4 3 2 1

To our husbands, with love

CONTENTS

CONTENTS

PART FOUR: LEGACY

LOVE WINS

PROLOGUE

JULY 16, 2013

SOON IT would be time for good-bye. His husband was dying, and with a gentle knock on the bedroom door long past midnight, it would come now, quickly. But Jim Obergefell, married for five days, didn't want to think about a funeral, not on this bright July morning when Cincinnati was in the throes of summer and his husband, John, was sitting up in bed because the spasms that coursed from his hips to his toes had mercifully subsided.

John Arthur couldn't wear a wedding ring. The weight of it hurt his fingers. He was naked under an electric blanket because clothing made his skin burn. His voice, what was left of it, had become winded and hoarse, a labored delivery of syllables and sounds that required great concentration and long, shallow breaths. Jim had to bend low to hear him, endlessly struck that a man who'd once had such a deep and lyrical laugh could now produce only a whisper. But for five days, John had pushed out a single, perfect word.

Husband.

Good night, husband. Good morning, husband. I love you, husband.

Disease had struck suddenly, just after John's forty-fifth birthday two years earlier, when his left foot started dragging as if a ten-pound weight was bearing down on his shoe and everything they knew shifted and splintered. The diagnosis of ALS had been a death sentence: the neurological disorder attacks the nerve cells in the brain and spinal cord, eventually robbing every muscle in the body of movement, including the diaphragm, which facilitates air flow to the lungs. Amyotrophic lateral sclerosis literally suffocates its victims to death.

Jim glanced at John in the bedroom they'd once shared, painted pale yellow and dominated by a hospital bed that compressed and expanded beneath John's weight. Jim had moved to the guest room, but already this morning he had spent several hours in a chair by John's bedside, watching the news on a television set loud enough to overcome the constant whoosh of an oxygen generator that pumped air through a line looping over John's ears and up into his nostrils. The bedroom faced east, and Jim had opened the window blinds so John could feel the sun.

On this day, they were expecting a visitor.

Jim was nervous about meeting a civil rights lawyer who had spent the better part of thirty years suing the City of Cincinnati, but when Al Gerhardstein rapped on the door of their downtown condominium just after two P.M., his smile was benign and his graying sideburns were slightly disheveled, as if he had just come in from a run. He adjusted his wire-rimmed glasses, and when he shook Jim's hand, the embrace was firm and friendly.

Al followed Jim down the length of the hall to the bedroom, where John was propped up on pillows, waiting. Dropping his briefcase to the floor, Al barely glanced at the hulking hospital bed. His younger sister, one of six Gerhardstein siblings, was paralyzed by multiple sclerosis, and on Saturday mornings, Al sipped coffee by her bedside until her caregiver arrived.

He leaned forward and rested a light hand on John's shoulder. And then he said, "Tell me about your wedding."

"Saying 'I thee wed' was the most beautiful moment of my life," Jim said, looking at John, whose frail frame was hidden beneath the blanket, and remembering the lanky, grinning man with a mop of blond hair.

They had spent more than twenty years together in Cincinnati, spread across the hills and low ridges of the Ohio Valley, but had never felt compelled to marry because Ohio had banned same-sex marriage and the federal government didn't recognize the state-sanctioned marriages of gay couples anyway. But three weeks earlier, the U.S. Supreme Court had delivered an important win to the gay community, finding that same-sex couples married under state law deserved all the federal benefits that came with it, spanning health care, Social Security, veterans' assistance, housing, taxes. So Jim and John had traveled to Maryland to marry, each mile in a private plane fixed with medical equipment wearing on John's fragile body.

"He suffered," Jim told Al. "It's frustrating and hurtful to know that the person you love went through terrible pain and discomfort just to do something millions of others take for granted."

"I wanted us to be treated the same," John said slowly, each word a struggle. "And I want Jim to be legally taken care of after I die."

Al listened without taking notes. Once, twenty years earlier, when he had been in his early forties juggling three children and a shoestring law practice that operated on contingency, the voters of Cincinnati had changed the city's charter, permanently banning all laws that would protect the gay community from discrimination in areas like housing and employment. To Al, it was an arbitrary and hateful provision, and he sued in federal court. He spent nearly five years working without pay, and when the case was done, he questioned the city, the courts, and the application of law. He thought about shuttering his practice and taking up teaching, moving his family from Cincinnati.

He wasn't sure if he would ever take on another major gay rights case, but then *United States v. Windsor* on June 26 had struck down a key provision of the Defense of Marriage Act, which for more than fifteen years defined marriage solely as a union between one man and one woman. The law, Supreme Court justice Anthony Kennedy wrote in the opinion overturning it, told gay couples "that their otherwise valid marriages are unworthy of federal recognition."

Over late nights in a dusty office overlooking a bus stop and the federal courthouse, a poster of Rosa Parks in the lobby and hate mail tacked to a bulletin board in the kitchen, Al studied Ohio's ban on same-sex marriage, passed by a majority of voters in 2004. He discovered a striking inconsistency.

Ohio prohibited first cousins and minors from marrying, but would recognize the marriages if they were performed in another state. The "place of celebration" rule came from a long-standing legal principle known as *lex loci contractus*, the Latin term for "law of the place where the contract is made." But Ohio didn't apply that principle equally: out-of-state marriages among same-sex couples were excluded. If the federal government after the *Windsor* decision had to recognize same-sex marriages on an equal basis with heterosexual marriages, Al thought, shouldn't states like Ohio have to recognize them, too?

The finding seemed entirely existential until a mutual friend told Al about John, Jim, and their marriage in Maryland. On a hunch, Al hurried to his filing cabinet and dug out a death certificate saved from an old lawsuit. He scanned the document from the Ohio Department of Health and immediately found what he was looking for. "This is it," he whispered, and three days later, he went to meet John and Jim.

When Jim got up for water, Al followed him into the dining room and pulled out the death certificate. "I'm sure you haven't thought about this, because who thinks about a death certificate when you've just gotten married."

Al pointed to Section 10: Marital Status.

Al pointed to Section 11: Surviving Spouse.

Jim had never studied a death certificate before, and he followed Al's fingers across the page. In Ohio, Al explained, their marriage did not exist. John would be classified as single when he died, and the box where Jim expected to be named as the surviving spouse would be left blank. In death, they would be strangers.

"John's last official record as a person will be wrong," Al said gently.

Jim's head started hurting as he thought about John, who had traveled nearly five hundred miles to marry even though his bones ached and his skin burned. "I can't believe this," he said, wiping at tears with the back of his hand. "I just can't believe this."

Across the country, lawyers and advocacy groups were focused on a broader mission—delivering same-sex marriage to all fifty states. But Al saw a more narrow, tangible, and urgent need, one that had never been fully explored in an American courtroom. He asked Jim, "Do you want to talk about ways that you can fix this for you and John and make a difference for all gays here in Ohio?"

Jim wasn't sure if there was room in their lives for a federal lawsuit, but when he walked back down the hall and looked at John, whose wedding ring sat on a nightstand by the bed, there was no longer any question. Jim had no idea where the case might lead or how it would end, but with John's permission, Jim signed John's name and then his own to some papers that Al laid out in front of them. Alone in the fading light of the yellow bedroom, Jim and his husband launched the first-ever challenge to Ohio's ban on the recognition of same-sex marriage, enshrined in the state's constitution by 3.3 million voters.

PART ONE

LOVE

"I love you without knowing how, or when, or from where,

I love you simply, without problems or pride:

I love you in this way because I don't know any other way

of loving but this."

—Pablo Neruda, *100 Love Sonnets*

1

INDIGNATION

JUDGE TIMOTHY S. BLACK strode into his bustling courtroom, long black robe whipping around his legs, and glanced at the sea of faces that was looking at him, the judge who would decide whether the will of Ohio's voters trumped the wishes of a dying man. He looked at the lawyers, one for the City of Cincinnati, two for the State of Ohio, and three for the plaintiffs, and at the spectators who filled every row of the federal courtroom, with its mahogany paneling and plush maroon carpeting. But the judge's gaze settled on one man in particular, whose tired eyes were hidden behind brown-framed glasses.

Jim Obergefell sat perfectly still at the plaintiff's table. The walk from Al Gerhardstein's law office to Cincinnati's Potter Stewart United States Courthouse had seemed more like a mile than a single block, down the sidewalk, across East Fifth Street, past the bus stop, and Jim tried to control the quivers in his stomach by thinking about John, who would have admired the architecture in the elegant room if he weren't home in bed, unable to move much more than his fingers.

Judge Black had called for an emergency hearing on the muggy July afternoon to decide whether to require the State of Ohio to issue a death certificate listing John Arthur as a married man. The fledgling case was already drawing newspaper headlines, and in the hours leading up to the hearing, the fifty-nine-year-old judge paced alone in his hushed office on the eighth floor of the federal courthouse. Only three days earlier, his law clerk had rushed in, clutching Al Gerhardstein's lawsuit. "This is going to be a historic case," she said.

The judge knew that any ruling would be narrow in scope, approving or rejecting the words on a single death certificate, but his decision would set legal precedent, opening the door to a broader challenge to the state's ban on same-sex marriage.

Should one judge step in and upset the democratic process?

It was a colossal question, and he glanced across the room at pictures of his wife, Marnie, who had listened for hours to his blistering critiques of the Vietnam War when they met for the first time along the shores of southern Maine in the summer of 1972. Voter initiatives were a form of direct democracy that the judge considered fundamental. But a married man was about to die a single man, at least in the eyes of his state, and so the judge had decided to call for an emergency hearing, clearing his docket and feverishly researching case law to find a similar lawsuit in another city, some kind of precedent that would offer guidance and legal backing before he waded into the tumultuous world of gay rights. He found none.

He thought about his mentor, federal judge S. Arthur Spiegel, who had talked about judicial courage long after he sent ballplayer Pete Rose to prison for tax evasion or oversaw a landmark settlement involving a nuclear plant. Make a decision without looking over your shoulder, the elder judge had told Black.

Judge Black had a tremor in his hands that flared up when he was anxious, but his voice was deep and measured when he walked into the courtroom and looked down from the bench just after lunch on July 22, 2013.

"Good afternoon, ladies and gentlemen. We're here in the open courtroom on the record in the civil case of James Obergefell and John Arthur. . . ."

Jim straightened in his seat and looked at Judge Black, whose expression was inscrutable. It had been only six days since the first meeting with Al, barely enough time to process the idea that they had just launched a federal lawsuit against the State of Ohio. Before court, Jim had drafted a statement that described his marriage to John, but he wasn't entirely sure how the judge would react. Jim had come to talk about love, but he knew the state's lawyers would argue the law—in particular, twenty-four words in Article 15, Section 11 of the Ohio Constitution.

Only a union between one man and one woman may be a marriage valid in or recognized by this state and its political subdivisions.

Jim looked over at the two lawyers representing popular Ohio attorney general Mike DeWine, a former U.S. senator and prosecutor with eight children and nineteen grandchildren. His wife, Fran, baked fruit pies for annual ice cream socials and political rallies, and the couple had opened a school and food program in Haiti, named after their late daughter. As a Republican in Congress, DeWine had voted in favor of hate crime legislation in 2000 but voted against it two years later. In 2006, he led a push for a federal constitutional amendment that banned same-sex marriage. But his opposition in Jim's case, he argued, was rooted in democracy.

"I don't have, as attorney general, the luxury of deciding which parts of Ohio's constitution to defend," he would tell the *Toledo Blade*. "If the voters of Ohio 10 years ago had voted to allow gay marriage and that was in the Ohio Constitution I would defend that against attack. That is what the attorney general does."

To Jim, the argument seemed more like rhetoric than sound reasoning, particularly since the attorney generals of California, Illinois, and Pennsylvania had recently announced they would no longer defend same-sex marriage bans. John and Jim had spent two decades together, largely removed from the politics of gay rights, but Jim's blood boiled every time he looked at his dying husband. That morning before court, John must have sensed it, too. He whispered to Jim from his bed, his voice weak and wavering, "Go. Kick. Some. Ass."

Al Gerhardstein arranged his notes and stood up, struck, as he always was, by the elegance of law. The courtroom was the only place where Al's clients were equal to the powerful people they sued, where a poor black man could take on a police chief, a battered woman the domestic violence policies of an entire department. In court, the rights defined by the Constitution were not ideals but absolute guarantees that applied to everyone, no matter how disenfranchised.

Before court that morning, Al put on one of his favorite ties, with pictures of smiling children. He turned to Judge Black and said, "The evidence is going to show that these two men are in love, and they've been life partners for more than twenty years. James and John were recently married in Maryland. . . . Marriage[s] between same-sex couples are legal in Maryland . . . but their marriage is not recognized in Ohio. Why is that? Because Ohio doesn't recognize any marriages between same-sex couples."

". . . So your argument is not that Ohio has to authorize same-sex marriage in Ohio, but your argument is that Ohio has to recognize another state's law that permits same-sex marriage?" Judge Black asked.

Al knew from experience that social progress often came in fits and starts, painfully slow and unaffected by the passage of time, and so he had decided to wage a narrow fight focused solely on Ohio's refusal to recognize marriages rather than the right to marry itself. Standing before Judge Black, the only thing Al would ask for was a single death certificate that recognized the marriage of John Arthur and Jim Obergefell.

"That's exactly it, Judge," Al said. "...John Arthur is dying. The evidence is going to show that he has days, or, at most, weeks left to live. When he dies, the final record of his life in Ohio will be his death certificate. Unless this court acts, the death certificate will not reflect his marriage at all and will not reflect that his husband, James Obergefell, is his surviving spouse."

Al glanced behind him. "Plaintiff calls Mr. James Obergefell."

Jim walked quickly to the witness stand, raised a shaking hand, and swore to tell the truth. He was grateful for the familiar faces in the courtroom, which included John's favorite aunt, Paulette Roberts, a community theater actress with cropped gray hair and an infectious giggle.

Al walked to the witness stand and looked at Jim. "I would like you to tell the court, just briefly, how long you and John have been together and what the relationship has meant to the two of you."

"We've been together since December 31, 1992, and it's been my world," Jim said, hoping his voice sounded steady. "It's been my life. We've been in a committed relationship since that time. And in our eyes, we are married. Our families love us. Our families consider us married. Our families and friends treat us as a committed married couple."

"Now, you've been a couple and lived together for twenty years. Why is it so important to be married?"

"Well, I think that's the same for any couple who decides to get married, no matter how long they've been together. We want our country, our state, to recognize our relationship and to say, 'Yes. You matter. You were married. You have the rights, the benefits, and the responsibilities that go with that, just as any other couple.' With John near death, it was very important to us to have our relationship formalized and recognized by our government."

"I don't want to belabor this," Al said seconds later, "but for the purposes of the record, we do need to have some sense of how imminent, if you know, John's passing might be."

Jim paused and took a breath. "I would say days, maybe weeks if we're lucky. Last week, the RN with our hospice service pulled me aside after their visit with John to tell me I should start preparing because she believes the end is close."

From her seat in the middle of the courtroom, John's aunt started crying.

Al placed an Ohio death certificate in front of Jim. "Now, there's a box Number 10 where it says, 'Marital Status at Time of Death.' Did I read that correctly?"

"Yes. You did."

"And if this court does not act, how will that box be filled in?"

"Unmarried," Jim said.

" . . . And the next box says, 'Surviving Spouse's Name.' If this court doesn't act, how will that box be filled in?"

"It will remain blank," Jim said.

"How should it be filled in?" Al asked.

"James Obergefell. My name should be there," Jim said in a rush.

"So, if you would, just tell the court how the problem with the death certificate and recognition of your marriage by the State of Ohio harms you and John."

Jim pulled out his written statement. "I have to read this," he said, looking down at his notes, "otherwise, I probably will not get through it."

"That's okay," Al coaxed. "Go ahead."

Jim shifted in his chair and looked up at the judge. "Your Honor, during our twenty years together, John and I have taken care of each other during good times and bad, for richer and poorer, and in sickness and in health. For the past two years, I've had the honor of caring for him as ALS has stolen every ability from him. Rarely a day goes by that he doesn't apologize for what he feels he's done to me by getting sick. He is physically incapable of doing anything to thank me or assuage his feelings of guilt, and we all know that there are times when words aren't enough. We need to do something."

Jim paused. "What he wants is to die knowing that I will be legally cared for and recognized as his spouse after he is gone. That would give him peace, knowing he was able to care for me as his last thank-you. When I learned that John would forever be listed as unmarried on his death certificate, nor would my name be listed as his spouse, my heart broke. John's final record as a person and as a citizen of Ohio should reflect and respect our twenty-year relationship and legal marriage. Not to do so is hurtful, and it is hurtful for the rest of time."

Judge Black glanced at the other attorneys and said, "Adverse parties wish to inquire?"

Bridget Coontz, the attorney for the State of Ohio, said, "No, Your Honor."

Aaron Herzig, the attorney for the City of Cincinnati, said, "No, Your Honor."

"You can step down, sir," the judge told Jim.

Exactly 3,329,335 people in Ohio had voted to reject same-sex marriage in November 2004, embedding the law in the state's constitution. Ohio had been one of eleven states that put the issue on the ballot on the same day that November, and voters in every state had passed nearly identical measures.

It was a stinging defeat to those in the gay rights movement who had long considered marriage equality among the most fundamental goals. After the Hawaii Supreme Court in 1993 ruled that denying marriage licenses to same-sex couples was discriminatory, it seemed to many that real change was underway. Beyond the promise of love and commitment, sanctioned by the state, marriage would simplify practical matters about children, property, money, medical decisions, births, and deaths. But Congress responded with the Defense of Marriage Act, defining marriage as a union between one man and one woman at the federal level, and the voters of Hawaii in 1998 put an end to the possibility of marriage by passing a constitutional amendment that gave the legislature, not the courts, the power to decide the issue.

The pushback was so sweeping that some leaders in the gay rights movement wanted to drop the fight for marriage altogether and focus instead on other fronts, like laws that protected the gay community from housing or workplace discrimination. But in 2003, the Massachusetts Supreme Judicial Court backed the freedom to marry after years of intense debate and litigation, clearing the way for gay couples to exchange vows for the first time in American history. Conservative groups countered with eleven ballot initiatives in November 2004, and in Ohio in the nine years since, no one had challenged the ban or its nuances.

Until now. Jim stepped down from the witness stand and walked back to his seat at the plaintiff's table, wondering whether Judge Black would see fit to save his marriage. Jim's legs felt wobbly and the whole scene seemed surreal, and he thought again of John, who had loved Ohio and had called it home for most of his life and now faced death without the state's backing. In that moment, Jim ached for his husband, who had learned early on, when he was a lonely boy in the suburbs of Cincinnati, what it meant to be different.

2

IRRECONCILABLE DIFFERENCES

SUMMER CAME fierce and steamy to Cincinnati in 1976, but John Arthur escaped to the woods, where a creek was filled with bluegill and a nest of snakes lived in a moss-covered Frigidaire buried beneath the water. He spent long afternoons digging for rocks or worms or kicking off his sneakers to wade in the creek. The clay at the bottom felt smooth and cool as it squished between his toes, and he would grin at his younger brother Curtis and their cousin Keith, who splashed and danced around him.

There was peace here, in the long shadows of old oak trees. The prospect of sixth grade in a school with few friends terrified John, but in the woods with his brother and cousin, he would laugh out loud as he struggled up hills covered with a fine layer of sand, clinging to vines and tree roots to steady himself. Late one afternoon, the three boys decided to edge past the creek, deeper than they had gone before. All around him, John smelled wood and wet grass, and he breathed deeply, the Cincinnati summer long behind him.

Suddenly, Keith stopped and pointed. "Look!"

The woods spanned one of the steepest hills in the city, and Keith was standing on the ledge at the highest point, fifty feet above a desolate stretch of dirt and rocks.

"It's got to be a thousand feet down," Keith called, peering over the ledge.

"The thousand-foot drop," John said slowly.

About ten feet away, a massive tree had toppled and lay partly over the ledge, fixed to the side of the hill by its roots. Keith, who had decided several years earlier that he had been blessed with the skill set of the Incredible Hulk, scaled the trunk, then crawled on his stomach out over the ledge.

"Come on!" he called.

Curtis, two years younger than John, immediately followed. John hesitated. He was asthmatic, uncoordinated, and so thin that Curtis and Keith had taken to calling him "Spaghetti Noodle." But even at ten, John was learning how to overcome weakness with willfulness, a defiance that would startle his family and humble those who doubted him.

At summer camp, a truck tire had once been placed between two teams, and Keith and John were pitted against each other, each charged with pulling the tire back to their side. Keith's team whooped and hollered; he was older, stronger, and cockier than John. But John wrapped his skinny arms around the tire and dug his heels into the ground as Keith tugged, red-faced and sweaty. John's head snapped back and forth as Keith pulled and pushed, but John had a death grip on the tire and wouldn't let go. John knew he couldn't win, but he could hang on just long enough for the other boys to rush in and help. John's team captured the tire, the sweetest win in his young life.

"Come on," Keith yelled again from the top of the ledge in the woods. It was John's turn to scramble out onto the tree trunk.

Slowly, John lowered himself onto his stomach, grasped the

prickly trunk, and scooted out over the ledge. He looked down, and for one glorious moment he felt as if he were dangling from the top of the world, surrounded by air and space, cut off from home and connected to this hushed corner of the woods in some indelible way. The three boys would return to the tree on the thousand-foot drop again and again, daring each other to scramble out over the ledge, sometimes on their stomachs, sometimes standing straight up with their arms out for balance.

Looking back, Keith thought Barry Mulvaney and his brothers seemed like little more than punk kids with nothing to do at the tail end of summer. But when they charged John, Curtis, and Keith in the woods late that August, Barry and his brothers were giants, a terrifying, stomping furor of pumped fists and curse words.

It had started the day before, when Barry wandered over to Keith, knocked his pitching machine to the ground, and punched him in the face. In the woods with his brothers, the feud fresh and simmering, Barry spotted Keith and his cousins and yelled, "Let's get 'em!"

John, who hated to run, sucked in his breath and started sprinting, desperate to escape the woods and the shrieking boys who were chasing them. Curtis followed close behind. They ran past the creek and over the hills covered with sand. They ran along the dirt path all the way to Keith's apartment complex at the base of the woods, where they yelled for their uncle.

"We're being chased!" Curtis shouted as John, covered in mud and grass, struggled to catch his breath.

The two boys and Keith's father ran into the breezeway and found Keith stumbling out of the woods, his face bloodied and swollen. Barry and his brothers had pinned him down on the bank of the creek. Keith tried to grab a tree branch to fight back, but it was three against one, and the punches kept coming.

"They just attacked," he told his father, spitting blood. His eye was swollen shut.

John had never seen a man run as quickly as Keith's father that afternoon, particularly one who was nearly six foot four and three hundred pounds. John thought briefly of his own father and wondered whether his dad would have given chase.

"You sons of bitches," Keith's father bellowed at Barry and his brothers. "You wolf-packed my boy."

The color drained from Barry's face as he and his brothers bolted toward home. They didn't bother John, Curtis, or Keith again, but the woods never seemed quite the same.

John dreaded the short, cold days of winter and the mellow sunsets that forced him inside his family's trim white house on Wolfangle Road, to his homework and his chores and his father's fraternity paddle, one inch thick and solid wood, which Chester Arthur used on his sons with a navy man's precision.

Years later, John's mother, Marilyn, would say that Chester had turned on his oldest boy early, even before he learned to walk. John fidgeted in school and fumbled on the soccer field. He had been appointed goalie on Coach English's team only because he was two heads taller than anyone else and didn't care much for running. John didn't want to play, but his father had threatened, "You call the coach yourself and tell him you're a quitter."

Once, John had grown so embarrassed during a physical fitness test at school, he ran straight off the running track, stumbled home, and hid in a tree until the principal coaxed him down. He hated his asthma inhaler. He hated wearing Brittania jeans when everyone else was wearing Levi's. More than anything, John hated a word that he didn't understand.

"Fag," the kids hissed on the bus to school. John would stare straight ahead, fighting hot tears. "Fag," they said again.

If Chester Arthur suspected his son was gay, he never said a word. Chester had grown up just across the Ohio River in northern Kentucky, the son of a candy maker called Pop Arthur, whose peanut brittle had made him something of a local celebrity. Handsome and trim, Chester joined the service after high school and spent two years in Pensacola learning to fly planes, but there was no war to fight in 1960, so the naval cadet came home to study economics. A businessman comforted by the consistency of numbers, he used graph paper to draw up a list of chores for his two sons.

John was assigned the vacuuming.

As chores went, this one wasn't entirely wretched. John could lose himself in the humming of the machine, and the burnt-orange carpet in the living room was never really dirty anyway. He was careful to avoid his mother's antique pie safe, a six-foot cabinet near the bay window that held Chester's prized Marantz stereo system. John had watched his father open the doors to the cabinet with great flourish, his fingers lingering over the shiny knobs and switches. Once, he had played Arlo Guthrie's "Alice's Restaurant" for John's cousin Keith, and then went straight to the record store to buy Keith the album.

One winter night after dinner, John pulled the vacuum cleaner from the closet. The house smelled of fresh coffee and his mother's cigarettes, and John was happy because his aunt had come for a visit. Paulette, Chester's younger sister, was an insurance executive and community theater actress. Called "Tootie" for as long as anyone could remember, she would swing John around by his arms and dance to Carly Simon songs. "Again, Aunt Toot," John would cry. "Again."

Paulette had loved John like her own from the moment he was born, perhaps because she sensed that her brother found John a great disappointment. Chester's garage was crowded with camping gear and tennis rackets, but John had never shown any interest. Paulette doted on her anxious nephew, told him he was perfect, and their easy conversations gave John great comfort.

Paulette and Marilyn were in the kitchen sipping coffee when John started running the vacuum across the living room floor. His father came up behind him and John stiffened.

"You're not doing it right." Chester's voice was low and strained.

John glanced briefly at his father, fighting a bout of panic.

"You're not doing it right," Chester said again.

His father dropped to his hands and knees, pointing to lint on the carpet. "Look here! Look here!"

John pushed the vacuum. Chester followed on his heels. John pushed. His father pointed.

"You missed a spot! Look! Look!"

Paulette and Marilyn heard the shouting from the kitchen, and they tensed over their coffee. But it was 1976, and the nuances of child abuse were not yet part of mainstream discussion. Chester's barbs were written off to a bad temper, irreconcilable differences between father and son. As the yelling grew louder, Paulette finally looked at Marilyn. "What's wrong with him?" she asked.

"I don't know," Marilyn said, her eyes fixed to the kitchen table. Petite and blonde, Marilyn Arthur had been a headstrong student in the early 1960s at Christ Hospital School of Nursing in Cincinnati, but Chester insisted she stay home when their boys were born, so Marilyn had painted the walls of their four-bedroom house yellow and white, cooked chicken and biscuits and nursed the tulips and rosebushes in the backyard. She could never quite figure out how to confront her husband, perhaps because Chester had grown up in a religious family that believed prayer, rather than deep discussion, could mend even the most troubled relationships. Though Chester had shrugged off faith as an adult, he communicated in short, clipped sentences, and Marilyn figured it was wiser just to stay quiet.

By the time Paulette reached the living room, John's eyes were red and wet.

"There you go crying again." Chester smirked as he sat back on his heels behind John, arms folded across his chest.

Paulette bent down and touched Chester's arm. She knew the only way to stop her brother was to distract him, so she said in a rush, "Ches, what are you up to this weekend?" Chester turned to answer and John ran to his bedroom, where he liked to squeeze inside a tiny closet with a shuttered door that was three feet off the floor, a cubbyhole of sorts that had been cut into the wall. With a foot of space, there was just enough room for his Matchbox cars and Hardy Boys books.

Paulette left a few minutes later, fighting a headache and unsure what to do. Cincinnati was a conservative, religious city heavily populated by German and Irish immigrants. Interfering with her brother's children was unimaginable, so Paulette went home, hugged her son Keith, and tried not to think about John's stricken face.

Paulette would tell her nephew, "If you ever need me, I'll come get you. Call me, and I'll be there."

In the kitchen, the refrigerator was covered with graph paper. The straight lines and sharp edges appealed to Chester, and as John grew older, there was a chart to log grades, a chart to log chores, a chart to log visitors, to be filled out in triplicate. *Name. Length of stay. Nature of visit.*

Curtis, who would earn a master's degree in education, filled out the charts every Friday. He played tennis with his father and chatted about fishing trips. But John was different. He pushed back.

It seemed to John as if his father had always been angry, even when he was pouring concrete in the backyard so his sons could have a swing set or leading John's Indian Guide youth group on sweaty hikes along the banks of the Ohio River. Once, Chester spent an entire weekend building a pen for the family dog, a basset hound named Beauregard that John loved fiercely even though Beau dragged his fat belly through the grass and made John's eyes itch.

The father who hiked and built swing sets seemed starkly different from the man with the fraternity paddle, who once thwacked both boys for talking with the lights on when they were supposed to be sleeping. Keith had been watching the television series *Kung Fu* in the next room when he heard his uncle yell, "Bend down!" and his cousins start to yelp.

But by high school in the early 1980s, the paddle had been stashed in the attic, and John had become the irreverent focal point of an eclectic social circle. During his junior year, he had sat alone in the bleachers during pep rallies, feeling every bit an outsider as his classmates in maroon and gold chanted, "Open the door! Bring the Spartans on the floor!" Then everything changed.

In conservative Cincinnati, John started listening to Depeche Mode and streaking his blond hair. He wore Pierre Cardin cologne and Izod sweaters. He began to use his sharp wit to win over classmates and, carousing with friends in the 1978 Mercury Zephyr he bought with money from odd jobs, John became the consummate entertainer.

He was drinking warm beer on a friend's horse farm late on a Saturday night when someone suggested a game of "Truth or Drink." John sat cross-legged in the grass across from one of his best friends, Wendy Bailey, a pretty brunette who had shown up with her new boyfriend. "Have you ever gone all the way with her?" Wendy's boyfriend asked John in a tough-kid twang.

Wendy was too startled to respond, but John didn't hesitate. Humor, he knew, attracted friends and deflected criticism. He looked down at his feet, blinked rapidly, and shook his head back and forth. "Aww. I'm going to have to drink on that question."

Wendy's eyes widened. "John? John?"

"I'm sorry, Wendy. I'm just going to have to drink on that one."

John turned to his cousin Keith and delivered a well-practiced eye roll.

"Tell him the truth! Tell him!" Wendy said, frantic.

John just sipped his beer, looking at Wendy and her angry boyfriend. John would have gone on drinking except that Keith laughed and it was someone else's turn to ask a question.

John stayed out of the house as much as he could in high school, especially when he brought home a report card with mediocre grades and his father banned him from using electricity as punishment. John didn't mind sitting in the dark, but he begged Curtis to answer his phone calls. One afternoon, John came home from school and heard his parents arguing in the bedroom. It seemed their fighting was growing more intense, and John tried not to listen. Then he heard his name. "John," his father said slowly, "is the biggest disappointment of my life."

John kicked shut his bedroom door so he couldn't hear any more.

He left home two years later and enrolled in the University of Cincinnati. Halfway through college, he came back to the house on Wolfangle Road to collect his late grandmother's chairs and end tables, stashed in the basement, which he would use to furnish his first apartment. He had worried about going home, but Marilyn assured him that Chester would be out of the house when John got there.

John parked his car in the driveway, let himself in, and crept toward the basement. The yellow hallways were still, almost peaceful, as if a happy family had lived here all these years. He passed the tiny closet in the bedroom where he had stashed his game of Monopoly. He passed the kitchen where his mother sat and smoked. Then John heard a noise in the driveway and hurried outside. Chester shot out of his car, sprinted toward the house, and accused John of stealing furniture.

John dropped the table he was carrying and backed away. "It's fine. I'll just go and get some stuff from Goodwill."

He got into his car and sped off, empty-handed.

"It got really, really bad, Aunt Toot," he told Paulette a few weeks later. "I mean really, really bad."

Years later, Paulette would take Chester's fraternity paddle and throw it in the trash.

John never actually told his parents that he was gay. They found out in odd, disparate ways, which suited John fine, since a formal announcement would have seemed preposterous in a family that had always avoided discussions of substance. His mother found out when she walked into the bedroom of the apartment that John was renting near the University of Cincinnati and discovered her son sound asleep next to another man, who was six feet seven inches tall, with long legs that dangled off the edge of the double bed. She slipped out quietly and never asked John about it.

Looking back, it had always been clear that John was gay, and the discovery that he was dating men was really no discovery at all. The only person who needed confirmation was Chester, but he and Marilyn had divorced when John and Curtis were in college and John was no longer speaking to his father.

In 1988, just after John's twenty-third birthday, Curtis agreed to meet Chester for lunch. Curtis would later leave Cincinnati for teaching and client-services corporate jobs in Chicago, Washington, D.C., Los Angeles, Saudi Arabia, and Toronto, but he would talk to John nearly every day, exchanging witty words and phrases to downplay the dysfunction in their family. "How is your father today?" John would tease his brother. Curtis would reply, "My father is great. How is your father?"

But Curtis loved Chester, and during visits, they would talk about politics and bridge games. Over lunch, Chester must have suspected something because he looked at Curtis and said, "John's gay, right?"

"Yeah," Curtis said slowly. He studied his father's face, searching for signs of a reaction. But Chester sat perfectly still and said only, "Oh."

"Do you have a problem with that?"

"No," Chester said carefully. "I don't think so."

"Good. Because if you cut John out of your life, you're going to cut me out, too."

Curtis thought about telling his father more. John's turbulent childhood had given way to an unsteady life and a string of dead-end jobs, one month at a silk flower shop, the next at a video store, and he was spending late nights dressed in black at Cincinnati's gay bars. But Curtis decided it was best to say nothing.

One of John's favorite spots had fifty-cent drinks and Pet Shop Boys remixes, and late on a Wednesday in 1992 with his old fraternity brother, Kevin Babb, John staggered toward his car when it was time to go home. Babb tried to pull the keys away.

"Kevin Babb-u-lous," John said, lingering over the nickname that he had assigned his friend, who had a square jaw and auburn hair. "I'm fine."

John tripped and nearly tumbled face-first into his powder-blue Crown Victoria, which he had inherited from his late grandmother. Though the car was huge—John dubbed it the *Battlestar Galactica*—he had fallen instantly for the crushed-velvet seats and dashboard made of vinyl.

"Keys." Babb held out his hand.

For months, Babb had been staying close to his friend at bars and parties, concerned that John's tipsy antics would lead to real trouble. It was a difficult decision for the budding investment adviser, who, like other gay men in Cincinnati in the early 1990s, had considered leaving town for more liberal cities like Chicago or San Francisco. Cincinnati had always been a vibrant Midwest hub, the first in the country to establish a Jewish hospital, a full-time fire department, and a professional baseball team. Its chili parlors and tidy riverfront thrived, and downtown, modern buildings were linked by long stretches of indoor skywalks.

But sin, some believed, had settled in the city.

The word was being spread by a handful of activists and organizations, including the National Coalition Against Pornography and Citizens for Decent Literature, which had set up offices in town and called on the sheriff and local politicians for support. Temptation was run clear across the river to Kentucky; in Cincinnati, adult bookstores and movie theaters were forced to close.

The movement was heavily focused on the gay community, particularly young gay men, who whispered about arbitrary arrests of "queers" in parks and clubs. Police had charged two men with disorderly conduct for creating a "physically offensive condition" after they were spotted holding hands in a parked car. And the director of Cincinnati's Contemporary Arts Center was indicted for obscenity after refusing to take down an exhibit by famed photographer Robert Mapplethorpe, which included explicit images of men in sadomasochistic poses. There was talk about local doctors, dentists, landlords, and hair stylists who had refused service to gay men, fearing AIDS, and employers who were giving pink slips to gay employees. *GQ* magazine would declare Cincinnati the "Town Without Pity."

Though Babb worried, John barely took notice. In 1992, he had just turned twenty-six, had no real career path, and was living with five gay roommates in a house that his mother had bought for him. That spring, John had lunch with his friend Meb Wolfe, who, at five foot two and sporting a pixie haircut, had a shelf full of self-help books and an intense interest in finding a suitable boyfriend. She had grown close to John after college over meandering conversations about astrology and their joint debacles with unavailable men.

John looked at Wolfe and put his head in his hands.

"I'm never going to find anyone," he whispered. "Who's going to want to be with me anyway?"

3

THE BOY FROM SANDUSKY

JIM OBERGEFELL adjusted his sweater vest and tried not to gape at the man in the middle of a tangle of friends, hunched over a gin and tonic with a sly grin and head of blond hair. He overheard something about a master plan to introduce a girl named Meb to straight men, and now the night out at Uncle Woody's bar in the summer of 1992 had grown into a raucous affair.

Jim had come with Kevin Babb, who had traveled with Jim on a work-study college trip to Germany several years earlier. They had huddled on a frigid November night and watched the Berlin Wall fall, cheering wildly as the cranes came and thousands of people, clutching babies and flowers, poured into the West for a first whiff of freedom. Someone handed Jim a chisel, and he carved out a tiny piece of the crumbling wall, certain that he would never again find himself so close to the front end of history.

"Jim," Babb said, edging closer to the bar. "This is John Arthur."

John grinned at Babb, his old friend and fraternity brother, said hello to Jim, then turned away to challenge the woman next to him to a drinking contest. Jim watched quietly, sipping a beer. He had known for years that he was drawn to men, but being gay simply wasn't an option for a Catholic boy from a working-class family of German heritage, so he had asked a girl to marry him just after college and told himself, even after they broke up, that his handful of experiences with men were more fling than anything of substance.

Now Jim wasn't so sure. John was taller than most of his friends, and when he threw his head back and laughed, long and lively, it seemed to Jim as if the world slipped away—Babb, the strangers at the bar, nothing but static. Watching John, Jim felt his face grow warm.

Years later, Jim found out that his interest in John on that warm Friday night in 1992 hadn't gone unnoticed.

"Sweater-vest guy liked you."

John and Meb Wolfe were walking to the car just after midnight. "Seriously," John said. "He did not."

"It was the way he was looking at you."

"Seriously."

Jim Obergefell, with brown eyes and thick glasses, had always carried a look of genuine intensity, even when he was playing kickball or sorting a stamp collection in his family's narrow, colonial-style house in downtown Sandusky. The industrial town along the shores of Lake Erie wasn't as big as nearby Cleveland or Toledo, but in the early 1970s, the mills paid a decent wage and a gleaming J.C. Penney stood at city center. Jim would ride around town on a bike with a banana seat until dusk came and his mother called him home for dinner.

He was in middle school when he started riffling through the pages of an old Sears catalog, looking for pictures of men.

Jim knew with certainty that he was doing something wrong,

something that wasn't *normal*, and for a straight-A kid whose only serious transgression was skipping out on Father Hoover's Sunday church service, the idea made his stomach churn. So he tore out his favorite pictures, stashed them inside a tin coffee can in the basement playroom, and told himself that he would never look again. But even after he went to the freshman dance with Mona, the homecoming dance with Cindy, and the senior prom with Jennifer, he kept sneaking down to the basement for another look. Jim's family was large—he was the youngest of six—but he couldn't imagine finding words to explain what he was feeling when he didn't understand it himself.

He loved his father, but like many of Sandusky's fathers, Arthur Obergefell worked ten-hour shifts to secure a tenuous spot in the Midwest middle class. A stocky man with slicked black hair, Art had smashed up his left ankle in a dry-ice factory and lost parts of two fingers to a press in a paper plant. But he earned enough money to buy a five-bedroom house and take his kids on a trip to Washington, D.C., in the summer of '74, a pop-up camper hitched to the family's Dodge Coronet. From the top of the Washington Monument, eight-year-old Jim and his parents had looked out across the sprawling lawns of the National Mall, marveling at how the Capitol seemed so majestic even though the president had just resigned. While their parents took the elevator to the bottom, Jim and his brothers raced down nearly nine hundred steps.

"You nearly beat us," Art told his children, though he had been waiting for ten minutes.

Jim was closest to his mother, Mary, a librarian who taught her son how to bake rhubarb pies and iron shirts, first the collar, then the shoulders, then the cuffs. They spent long afternoons reading Agatha Christie novels and window-shopping at Sandusky's new mall. Mary had a sharp tongue but was soft on her youngest son, perhaps because he was number six, or maybe because she sensed that Jim craved the nurturing. She sent him to public school instead of Catholic school and slipped him money for new Polo by Ralph Lauren shirts and penny loafers. Jim would put a German pfennig into the slot meant for pennies. "Twerp," his older sister Ann teased, "you're a clotheshorse."

Once, home from college for Christmas in 1984, Jim considered confiding in his mother. A package had arrived at the front door, and his mother watched as Jim unwrapped the box and pulled out a steel-blue sweater. "Well, who sent it?" she asked.

She looked at the card and paused. "A boy?"

Jim wasn't sure what to say, and in the warmth of a living room that smelled of fresh butter cookies and pumpkin pie, he struggled not to cry. As a freshman at the University of Cincinnati, he had recently had his first experience with a man, another student down the hall in the dormitory. "Oh God," Jim thought afterward. "I'm going to die."

For weeks, he had been scouring his arms for red spots, certain that he would end up with the strange new disease that seemed to be inflicting gay men with a rare form of skin cancer. Every time he heard another news report about the rapid spread of an illness that was originally dubbed GRID, Gay-Related Immune Deficiency, Jim fought bouts of pure panic.

How could he share his thoughts about death and disease with his mother, whose cream-and-yellow kitchen was stocked with collectible Snoopy drinking glasses? How could he explain that the boy from the dorm who, just briefly, had made him feel so happy and alive had now bought him a chunky-knit sweater and sent it all this way to Sandusky?

His mother was watching closely, so Jim said quickly, "Oh, we're just really good friends."

The lie hung in the room. Jim would always wonder what it might have been like to tell his mother about his feelings, the sublime rush of adrenaline and attraction, the shame and confusion that lingered afterward, the worst kind of hangover. As it happened, he never got the chance.

Three months later, while he was singing with the university's chorus at a concert in Philadelphia, a call came from his sister. Ann was sixteen years older than Jim, a Catholic-school librarian who had three children of her own.

"You have to come home." She could barely get the words out. "Mom's in the hospital."

"What happened?"

"We've got to get you home."

With the university's help, Jim caught a flight to Pittsburgh and then to Cleveland. Every minute in the air was agony, and he was sobbing by the time he stumbled off the plane and embraced his sister at the airport.

"How is she?"

"She's still with us," Ann said. "That's all I can tell you."

Ann desperately hoped that was true, but in fact she had no idea what they would find at the hospital, where their mother was having emergency surgery. It had started with a painful red blotch on Mary's abdomen, which grew bigger just before she was supposed to leave with Art for a first trip to Disney World to celebrate their thirty-fifth wedding anniversary.

"Don't you think you ought to go to the doctor?" Ann had asked her mother, a lifelong smoker.

"No, because I'm going to Florida tomorrow."

"Mom, what if you die down there?"

Mary shot back, "Put me in a glass coffin so I can look out to see what I missed."

They were in Florida for only a day when the pain spread and doctors ordered Mary to fly home. Just before surgery, Ann told her mother that she was going to call Jim.

"Don't you dare," Mary said. "It won't matter because if I'm going to die, I'm going to die. Him being here and missing out with his friends and his choir is not going to change anything."

In the waiting room, Ann told her husband, "He's going to hate us forever if we don't call."

Mary was barely conscious when Ann and Jim arrived at the hospital. Though doctors had operated, an infection from bypass surgery years earlier had coursed through her body.

"Mom, Jim's here."

Mary gave a weak smile and mouthed, "I told you not to call."

Jim grasped his mother's hand. "I'm home, Mom."

She died four hours later, three months shy of her fifty-seventh birthday. Jim wanted to stay in Sandusky with his family, but his father urged him back to school.

"Mom would be pissed if you don't go back," Art told his son.

Five years later, his secret still in tow, Jim became the first in his family to earn a college degree.

Jim winced at the words he was sure every guilty gay boy or girl in America had uttered at one time or another. *There's something I need to tell you, and I don't know how you're going to take it.* In 1992, seven years after his mother died, he was that boy, alone with his father on their front porch in Sandusky. The house was bizarrely quiet, and Jim wondered whether he would ever get used to his father living alone.

Jim breathed deeply. "Dad. I'm gay."

There. It was only the second time in his life that he had said the words out loud. The first time had come several weeks earlier at graduate school at Bowling Green State University, where Jim was studying to become a student affairs administrator. On a weekend road trip, one of Jim's friends had asked, "Are you gay or straight?"

Jim, who had just met John Arthur at Uncle Woody's bar, readied a practiced denial. But from the backseat, he surprised himself and said simply, "Gay."

He wasn't sure if it had something to do with the insulated world of graduate school, or if, after years of teetering on the lonely edges of the gay community, he finally decided it was time to jump. But he raced home to Sandusky, crazy loose and incomprehensibly happy, because he couldn't stand to wait another day to tell his family. Now as he looked at his father, sitting stiffly on a folding chair on a humid summer night, Jim nearly lost his nerve.

But Art Obergefell said, "All I've ever wanted was for you to be happy."

Jim, too surprised to speak, hugged his father. "Thank you, Dad."

The next day, he was that boy again, meeting his sister for drinks at TGI Fridays. Jim had asked her to come alone. "Annie, I have something to tell you."

She grabbed his hands from across the table. "What? You're gay?"

"Really?"

"Yeah, well, we just figured it out," she said. "But you're still Jim, and you're still part of our family."

"Dad told me that, too."

"Well, what are we supposed to do? Turn it off and say we don't love you?"

"That's what some people do."

"Not in this family," she said.

By the time Kevin Babb invited Jim to a New Year's Eve party later that year, Jim felt as if his journey as a young gay man was just beginning. Babb's friend John Arthur was hosting the party at a house he had nicknamed the "Abbey" for its stained-glass windows and yellow-and-green trim. He shared it with five men, including Babb, and the party promised to be a boisterous affair.

Jim wanted to see John again. They had talked briefly for a second time at Uncle Woody's bar, several months after they first met there. Instead of a sweater vest, Jim had worn a snug T-shirt with the sleeves rolled up. John looked him up and down and said, "You'd never go out with a guy like me."

Jim was feeling bold, so he responded, "How do you know? You've never asked."

John never did.

Now the party at the Abbey was in full throttle, with a hundred people, friends from the clubs, friends from John's newest job in customer service, friends from college and the fraternity, swaying to ABBA songs on the living room's hardwood floors. Jim found Babb and they laughed as one of John's coworkers pulled a chair straight up to the dining room table and ate several dozen shrimp in quick succession. "Shrimp sucker!" John said, grinning. John wandered into the kitchen, and later, Jim found him surrounded by friends, drinking Tanqueray gin and leaning casually against the Formica countertops with his hands in the pockets of his jeans.

"You're handsome," John said when Jim walked into the kitchen. "I want to talk to you."

Jim thought he had never known another gay man who looked so completely comfortable with himself, which could have been why he walked straight up to John without thinking, slipped behind him, and casually put his hands in John's pockets. They stood there for what seemed like hours, talking to friends and waiting for midnight. Then John, who had been drinking much of the night, started to stagger.

"Let's go take a shower," he said to Jim, pulling him toward his bedroom on the third floor, in a converted attic with sloped ceilings.

The invitation wasn't sexual. Jim could see that John had had too much gin, so he quickly took charge upstairs, cleaning him up and settling him into bed with a wet head. Later they would learn that one of John's coworkers had opened the door to the bathroom, stumbled out, and nearly collided with Meb Wolfe and Kevin Babb at the bottom of the stairs. "Um. I think there are two guys in the shower," he said with a nervous laugh.

Wolfe and Babb headed upstairs to find John and Jim in the bedroom. Someone downstairs yelled, "We made it through another year," and in the first seconds of 1993, Jim turned to John and gave him a light kiss. "Happy new year," Jim said.

Kevin Babb had always teased John, "You'd be a real handful to date."

But after the New Year's Eve party, Babb watched his impulsive friend burrow on the couch next to Jim, who had stayed over the first night, then the second, then the third, wearing John's clothes and a wide, happy grin. Since John's boyfriends had often overlapped—John had nicknamed one of them "Wednesday"—Babb wasn't a bit surprised when, in the first few days of 1993, friends asked, "Is Jim still there?"

"Oh yeah," Babb said. "Jim's still there."

Jim didn't want to be anywhere else. He loved the way John charmed strangers at restaurants and called his female friends "dear." Jim had never met anyone with such a sharp wit, with one-liners about his broken family that made Jim laugh or cringe or both.

"Wait until you meet my mother," John said over lunch on the second day.

Jim thought briefly about his own late mother, who had worn wool suits and prim pink blouses, then nearly spit up his Diet Coke when John deadpanned, "She bought me this house so that I wasn't in the way of her sex life."

"You've got to be kidding," Jim said, trying not to stare at John's slender nose, which Jim found sexy. "There's no way."

John described his father, who had once blamed him for damaging the transmission on the family car. "I had to write a report," John quipped, "on how standard transmissions work." Jim imagined John as a teenage boy, and, just briefly, saw a hint of pain behind the smirk on John's face.

The fourth day came far too quickly, on an icy morning in early January when it seemed all across Cincinnati, Christmas trees were coming down and crumpled wrapping paper choked the trashcans at the curb. Jim was due back to graduate school at Bowling Green, three hours north, but the campus suddenly seemed far-flung and unfamiliar. He wanted to stay in the Abbey with John, watching *Soapdish* and sneaking kisses in the kitchen. But Jim had an internship, and he knew he needed to get back.

Over lunch on Jim's last day in Cincinnati, he took a slow breath and asked John, "Where do you think this is going? There's something here between us."

John hesitated and looked at the ground. "You know, I'm not much good at relationships."

"I know that."

"I've dated a lot."

"Yes. I know."

Jim started to feel queasy. The fidgeting man sitting across from him seemed entirely unsure about the possibility of a serious relationship, and Jim, whose parents had been married for thirty-five years, couldn't fathom anything else.

"John?"

John chose his words carefully. "I actually don't expect to live past thirty."

Jim knew about the asthma, the bars, the parties. He wanted

to reach across the table and touch John's hand, but he held back. John's eyes were expressionless, distant. "Don't say that."

"I'm serious," John said.

"We'll be okay. Can't we just see where this goes?"

Two hours later, Jim left for school, a newly out gay man newly in love, facing a gloominess that he couldn't quite imagine, a life outside the Abbey, without John.

4

ON THE MATTER OF FAMILY

IT SEEMED to Jim as if life in the first seven weeks of 1993 was a blur of calendars and countdowns, the exquisite anticipation on Friday afternoons when John would come visit, the lonely ache on Sundays after another long good-bye. Jim trudged through the four days in between, trying not to think about John's warning over New Year's weekend. *I'm not much good at relationships.* By some miracle, he seemed willing to try.

It wasn't a miracle at all, really, more of a don't-be-a-moron pitch from Meb Wolfe, one of John's closest friends. "Why can't you understand that somebody actually cares enough to want to be with you?" she had demanded.

John tensed. "I'm still processing it."

"Why?"

"What if he leaves?"

Wolfe sighed. She often dragged John along as her date to parties and weddings, where he analyzed her topsy-turvy love life and told her she was crazy when she said, "I must have a fatal flaw that everybody in the world can see but me."

John would gently reply, "Maybe you're looking in the wrong places."

Wolfe, in turn, had decided to help John kick his habit of attracting and then discarding a string of boyfriends. She had long known that her sensitive, skittish friend needed something in the way of unconditional acceptance to kick the insecurities forged in a home with a distant father, and Jim was offering a steady, stable hand. She looked at John and pleaded, "You need to give this a chance."

John looked terrified. "Give this a chance," Wolfe said again.

On a frigid Friday seven weeks after New Year's, John drove along rain-slicked highways to meet Jim at a higher-education conference in Columbus. Unpacking in Jim's hotel room at dusk, cold rain lashing at the windows, John pulled out a small box, neatly wrapped in green-and-black paper. "I got you something."

"What?" Jim said, flustered. "I don't have anything for you."

John moved closer. "Go ahead. Open it."

Carefully, Jim pulled off the wrapping paper and lifted the lid on the black velvet box. Inside was a gold ring with five diamonds set side by side. "Holy shit," he breathed, looking up at John and then slipping the ring on his right hand. He had never worn a ring before, and the metal felt cold and snug against his finger. John pointed to his own ring, a diamond solitaire that he had been wearing for years.

"See?" John said. "They're different but still the same."

Jim couldn't ever remember a feeling quite as pure as the one that swept through him at that moment, and he breathed deeply, silly-happy and utterly certain that the 180 miles between graduate school and John was a distance he could no longer tolerate. He needed to be in Cincinnati.

"Come over next weekend," John said, pulling Jim close. "It's time you met Mother."

Marilyn Arthur, who had navigated a broken marriage with painful restraint, made it a habit after the divorce to cuss, vent, judge, and otherwise speak her mind. "I only went off the pill twice," she frequently said, pointing to John and Curtis. "You were a sponge baby and you were a foam." She shed her A-line skirts for rubies and fur coats, took on two favorite new words, *fabulous* and *glamorous*, and, with money inherited from her parents, bought John a three-bedroom house. She upgraded to the six-bedroom Abbey after John convinced her that he needed something in the way of a formal entrance. Marilyn quickly became a commanding fixture in the house, mingling with John's five roommates, who promised to pay rent but rarely did.

The prospect of meeting Marilyn terrified Jim, but he agreed to a visit.

Jim was watching television with John's roommates early the following Saturday night, the new ring on his right hand, when he heard footsteps on the porch. Marilyn had come to get John for an outing to the Cincinnati Ballet. It had become a comfortable routine for mother and son that often ended with a late-night cocktail at one of the city's gay bars, where Marilyn held court and showed off her jewelry. More than once, John watched his mother shimmy and shake amid a throbbing cluster of dancing men.

John had planned to introduce Jim to his mother before the ballet, but John was still upstairs dressing when Marilyn arrived. Jim started for the door. "Don't," one of John's roommates called from the couch without looking up. "She has her own key."

The doorbell rang insistently, and from inside the living room, Jim could hear Marilyn bellow, "Get off your fat ass and open the goddamn door."

Jim nearly choked. He sprinted for the door and opened it, briefly wondering how such a pretty, petite woman could yell with such gusto. Marilyn looked him up and down for what seemed like an excruciating minute. "You must be Jim," she said, brushing past him to find John. "Not off to a very good start, now, are we?"

Jim could feel his face turn red. "They told me you had a key."

It wasn't long after that when Marilyn brought Jim into the fold. "I have three sons," she told her friends, and no one dared to question her. That spring, she offered to pay Jim's tuition at graduate school, but Jim decided to leave school altogether and move into the Abbey with John. Jim got a job in customer service, and when his former German teacher from the University of Cincinnati said she needed a good tenor for the church choir just across the river in Kentucky, Jim quickly signed up. He took John to meet Jennifer Kelley, a striking, rather formal woman who played Renaissance and Baroque music on the harpsichord.

Months earlier, Jim, feeling like an anxious boy again, had told Kelley that he was gay. "Oh sweetie," she said, each word slow and deliberate. "The only thing that concerns me is that you have been worried and unhappy."

Now Kelley looked at Jim, who was looking at John, and thought: you two belong together. She tried to convince John to join the choir, too, but John folded his arms, sat down in the pews, and said, "Nope. I am the audience."

John and Jim visited John's aunt Paulette, who had retired from her job at a Cincinnati insurance company and moved with her husband to a hundred-year-old farmhouse in the mountains of West Virginia. While four golden retrievers with muddy paws chased rabbits through the hillside, John and Jim sipped red wine on the front porch, swatting at flies and listening to Aunt Tootie's stories about life in Appalachia. Jim hovered over John, worried about his allergies and asthma, and Paulette thought she had never seen a young couple look so content.

John didn't like Christmas because it dredged up memories of being forced by his father to decorate a tree even though he was allergic and would often head to bed with a rash on his arms. But that first Christmas with Jim, John decided to visit Chester, who had met and married another woman named Marilyn, with five children of her own. John called her Marilyn 2, and without her, he may never have spoken to his father again. She had refused to marry Chester unless John was invited to the wedding.

Jim was entirely unsure what to expect from a navy man who would be meeting his son's boyfriend for the first time. In the car, John said, "Now, before dinner, there will be cocktails. There will be lots of wine with dinner. And then Dad will probably drink crème de menthe over ice after that."

Jim found Chester well-spoken and intelligent and was momentarily silenced by the similarities between father and son. But by the time port was served with dessert, Jim had stopped thinking altogether, and the last thing he remembered just before John hauled him home was the blurry image of Marilyn 2 pushing him toward the powder room. "It's the smallest room to spin," she said, dragging a cold washcloth across his forehead.

The only thing that made Jim unhappy in those first happy months with John was that his mother wasn't there to see his emerging life, not just life as a gay man, but life as an adult with a job at a financial services firm. What would she have seen when she looked at him, no longer a closeted college freshman but a grown man who had found love? The question made Jim ache.

But he grew closer to Marilyn, who spent long afternoons at the Abbey. On John's twenty-eighth birthday that year, she watched quietly as Jim set out to make an elaborate jelly roll dessert, with layers of whipped cream and strawberries rolled into chocolate cake. But the warm cake crumbled in his hands. "It's a disaster," he said, his fingers covered with crumbs.

"That's all right, honey," Marilyn said, pushing Jim aside with her hip and grabbing a spoon. "Let's just make a trifle."

Jim had never heard of the English dessert, but Marilyn calmly scooped up the chunks of cake, spooned it into bowls, and slathered the cream on top. Later, John would tell Jim, "That's when I knew Mother liked you. If that had happened to any of my other boyfriends, she would have just watched them fail."

It seemed to those who knew them that John and Jim had a yin-yang connection, diametrically different in a harmonious way, and more than once, friends in unhappy relationships would remark over a last glass of wine, "Will I ever have that?" John and Jim rarely fought in those early years, mostly because Jim quickly realized that it was impossible to stay mad at a radiant man who didn't seem to get what had gone wrong, like a happy toddler caught playing with a purple crayon on a white couch.

Once, Jim hurried over to John's desk at work so they could catch an evening flight to Paris. Jim hated to be late—he considered punctuality a trait passed on from his German ancestors—and when he found John still at his desk, absently chatting by phone with a client and nearly buried under a precarious stack of paper, Jim panicked.

When John hung up, Jim urged, "Come on, John. We've got to go. We've got to go." Jim was growing more frustrated by the minute, worried that they'd miss their flight.

John smiled and started sorting through the piles on his desk. Jim raised his voice and said, "You knew we were leaving. Stop!"

"Tone," John said, wagging a finger at Jim. Disagreements of any kind made John uncomfortable, and he often used that word to signal to Jim that irritation had crept into his voice. "Just one more thing and I'll be done," John said calmly.

They drove to the airport and caught their flight. "See?" John said, grinning, as Jim slumped in his seat, exhausted. "We made it."

Jim felt that same sense of exasperation when he arrived home from a trip to see family on a warm spring night in 1994, tired from traveling and anxious to see John.

"How did it go, Boo?" John asked, hugging Jim at the front door. "Did you have a nice time? Oh, and I made an offer on a house."

Jim let his suitcase fall to the floor. "Huh?"

"I put in an offer on a house." John was shifting from foot to foot, giddy.

They had been house hunting for weeks, musing about the perfect layout, the friendliest neighborhood, the benefits of a fixer-upper or a renovated home. But Jim couldn't imagine John making such an important decision without him, on a whim, while Jim was away with family. Maybe it was a joke, John being silly.

But John said, "I know you're going to love it."

Jim looked squarely at John, furious. "What the hell?"

"You're going to love it," John said again.

"John?" Jim's voice was rising.

"It's a cottage with two bedrooms," John continued, ignoring Jim. "Trust me."

As it turned out, Jim did like the house, but the deal fell apart over a bad inspection, and they ended up buying a Cape Cod in Cincinnati's hilltop Mount Washington neighborhood. After ripping out the wood paneling and refinishing the oak floors, they put a koi pond out back and settled right in.

Cincinnati was in the throes of a dusty, muggy spring in 1995, the worst kind of wheezing weather for a lifelong asthmatic, and John had been sick for weeks. But the cry from the bathroom came so suddenly that Jim had to stop for a moment to be sure he wasn't hearing things. Moments before, they had been sitting on the couch watching television.

There it was again, barely a whimper. "Jim?"

Jim raced toward the bathroom, stumbling over furniture. Something was wrong. He could tell by the raspy sound of John's voice. Where was that damned inhaler? Jim couldn't remember. He flung open the door and found John holding on to the edge of the countertop so tightly that his knuckles were white. His jeans were wrapped around his ankles.

John gasped. "Call 911."

Jim found the phone, but for a brief, horrifying moment, his fingers wouldn't dial. Finally, he spit out some words. "Allergic asthmatic. Can't breathe. Please. Come." He called Marilyn, who had just been to the ballet, then rushed back into the bathroom to pull up John's pants. It seemed as if John had stopped breathing altogether, and by the time Marilyn burst into the house minutes later—she had chased the fire truck down the street—Jim couldn't speak. He pointed. He sobbed. He wandered through the kitchen and living room, certain that his twenty-nine-year-old partner, who had once told his dubious friends that he would die before he turned thirty, would get them all good, right on time.

The paramedics squeezed into the bathroom around John, who was curled up on the floor. "John? Can you hear us?" one of them shouted. Jim watched as they lifted John onto a stretcher, unconscious, and loaded him into the ambulance beside Marilyn, who had gone back into nursing after her divorce. Curtis was there, though Jim couldn't remember when he'd shown up. Curtis coaxed Jim into the car and they followed the ambulance to the hospital. John was rushed into intensive care, and Jim, Marilyn, and Curtis spent an agonizing hour in the waiting room.

What if he died here on Mother's Day weekend? Jim thought

miserably, in the dismal green hallways of the emergency room. They were supposed to have been on a boat on the Ohio River, celebrating a friend's wedding. Jim had urged John to stay home just in case John's asthma flared up, but "just in case" was never supposed to happen, not in their air-conditioned house after a movie night on the couch. Now Jim, too, struggled to catch his breath.

Finally, a doctor approached. "We have him breathing and stabilized." A violent reaction to new asthma medication had restricted John's airways, and he had gone into respiratory arrest. The words sounded so ominous, Jim wanted to scream. "Do you want to go back and see him?" the doctor asked.

Jim hesitated, tears streaming down his face.

They were in a Catholic hospital in Cincinnati, where it wasn't good to be gay. The city's voters two years earlier had not only repealed a law meant to protect the gay community from discrimination but also amended Cincinnati's charter to specifically ban any legal protections moving forward, one of the harshest laws of its kind in the country.

In the days leading up to the passage of Issue 3, John and Jim had heard radio commercials that declared, "The homosexuals already have equal rights. They're asking for special rights and that's not right. Let's stop this in Cincinnati."

A group called Citizens United for the Preservation of Civil Rights had released *Inside the Homosexual Agenda*, a video that flashed pictures of writhing, kissing gay men in leather, bras, or heavy makeup. The narrator warned of promiscuity and child molestation in an AIDS-stricken gay community. "What is at stake is the future of America," U.S. senator Trent Lott declared on the video.

Standing in the hospital waiting room, Jim feared long looks and odd stares for the first time in his three years as a gay man in a long-term relationship. A nurse had asked him who he was. He wasn't immediate family, but John *was* his family. What if he was told that he couldn't go back to be with John? It was a gut-wrenching possibility that would stick with Jim for years.

Fully expecting to be told to wait outside, Jim said, "I'm his significant other."

The nurse paused, but to Jim's immense relief, she showed him to John's room. He was unconscious with a breathing tube snaked down his throat. Jim took John's hand and said, "You need to come home."

For two more days, Jim crouched by John's bedside in the hospital, waiting for him to wake up and unsure if he ever would. The thought of life without John was nothing less than terrifying, and Jim paced the room and the hallways, waited some more, unable to sleep or eat. Finally, late on a Monday night, John opened his eyes. He tried to talk, but the breathing tube was still in his throat.

"Don't," Jim said, rubbing John's forehead.

When the doctor took the tube out, John looked at Jim and whispered, "Good thing we didn't go to that wedding."

Jim smiled for the first time in two days, thinking about the petty annoyances that had inched into their relationship. John considered bad reality television a weekend sport. He left messy piles of mail and magazines on nearly every flat surface in their house. He was impulsive and quick to spend money, once heading to the home improvement store to a buy a wood slat for their bed and returning with design plans for an entirely new kitchen. But as Jim looked at John in the hospital, groggy from the near fatal asthma attack, he no longer cared about anything but getting John home. Despite John's ominous predictions, he would live to see his thirtieth birthday.

Jim bent low and whispered, "I hope nothing like this ever happens again."

PART TWO

LAW

"There may be times when we are powerless to prevent injustice, but there must never be a time when we fail to protest."

—Elie Wiesel

5

THE CHICKEN FARMER'S SON

TWO ABORTION clinics had already burned and an angry swell of picketers had descended on a third, hastily set up in a squat, red brick building along one of the busiest streets in Cincinnati. Al Gerhardstein was on edge, worried about late-night bomb threats and the man who seemed to be in charge of it all, a mail handler from Hebron, Kentucky, who wore ski masks during protests and described himself as the "defender of the unborn."

On a frigid February morning in 1987, Al walked to the courthouse to see a judge about prosecuting picketers who had rushed the clinic's doors, chanting "Damned to hell!" and elbowing the young women who struggled to get by. As the lawyer for Cincinnati's Planned Parenthood, Al had successfully pushed for a court order to keep picketers across the street. But every Saturday morning, dozens defied the injunction, bounding off the sidewalks and falling to their knees at the building's front doors. As volunteers shielded patients, Al took names on a four-page contempt of court form, which he rushed straight over to a court clerk for a signature.

It seemed to Al as if Cincinnati had become ground zero for the national extremist groups that were using kerosene and sledgehammers to destroy clinics. Two years earlier, someone had tossed firebombs into the basement windows of two of the city's women's health centers, causing hundreds of thousands of dollars in damages. Planned Parenthood set up a temporary clinic just down the street from the Cincinnati Zoo, but picketers gathered there, too.

Al had stationed video cameras on the roofs of nearby buildings. Now he walked quickly through the courthouse with footage of the most recent protest in his worn leather briefcase.

The thirty-five-year-old civil rights lawyer would see Judge Thomas Crush, a well-respected Republican who had issued strict orders to rein in the picketers. Already the judge had thrown several dozen people in jail for contempt of court, and Al had come with a list of new names. The U.S. Department of Justice would eventually declare anti-abortion violence a form of domestic terrorism, but back in 1987, the attacks on clinics in states like Virginia, Alabama, and Florida had only begun to intensify.

Al was organizing his files on a desk at the front of the courtroom when his cell phone rang. He had recently started carrying a portable phone, and it was the size of a brick, gray and bulky. The sound startled him, and he quickly whispered, "Hello?"

"Al, there's a suspicious package next to the clinic."

Al pushed his chair back and stood up. Ann Mitchell, the executive director of the city's Planned Parenthood Association, sounded unusually panicked even though she had been fielding threats for months.

"We've evacuated the clinic," she said in a rush.

When the judge came in minutes later, Al said, "Your Honor, we need to stop for the day."

He drove straight across town to the clinic, terrified that someone would end up hurt. The firebombs that had torched the health centers two years earlier had been planted when the buildings were empty. A daytime explosion at a clinic packed with doctors and patients would surely be lethal. The bomb squad had already cordoned off the streets when Al drove up. Swearing, he turned around and headed back to his office in the historic, fourteen-story Cincinnati Enquirer building, one of the city's most iconic high-rises. Al called Mitchell, trying not to think about the abortion doctor from Illinois who had been kidnapped or the clinic in Florida that had been bombed in the early-morning hours of Christmas three years earlier.

"It's okay," Mitchell said, sounding exhausted, when she finally answered the phone. A bomb, six inches long with a cigarette for a fuse, had been found leaning against the wall of the clinic. The fuse had been lit, but the bomb squad dismantled it in time.

Al shook his head, his fingers wrapped tightly around the phone. This has just escalated, he thought.

Had he known as a boy that he would one day defend abortion clinics, Alphonse Gerhardstein probably would have been stunned. In the late 1950s, the Roman Catholic priests at St. Mary's Parish in suburban Cleveland had seemed almost otherworldly, and Al, an altar boy, spent much of his childhood trying not to fidget through endless Sunday church services with his four brothers and sister.

Al had inherited his mother's olive skin and angular features. Carolyn Gerhardstein was a second-generation Italian who cooked boundless batches of rigatoni and volunteered at the parish, ironing robes and dusting the shelves of the sacristy. She worked nights as a nurse at the local hospital, and though the family wasn't poor, they worked a large garden to put extra food on the table and canned fruit from the cherry trees and currant bushes behind their wood-frame house on the outskirts of Cleveland. Al's father plowed a makeshift baseball field into the backyard.

Richard Gerhardstein, stocky and a good foot taller than his wife, had been a medic during World War II but took a job driving dairy trucks after he left the service. There was no money for vacations, but on Saturday mornings, Al and his father would drive to a market on the west side of town to pick up day-old bread and food in dented tin cans that were considered unfit for selling. Opening one of the unlabeled tins was like playing the lottery to eight-year-old Al, who never knew whether he would uncover green beans or potatoes or dog food. On a lucky day, he would find a fruit cocktail, one of his favorite desserts.

In the summer before sixth grade, Al's father won a promotion and the family moved to Parkman, a rural corner of Ohio forty miles east of Cleveland. Parkman was more of an outpost than a town, with a one-room post office, an aging grammar school, and a small Catholic church. Amish families traveled the narrow streets in horse-drawn buggies.

Al's father had been named manager of a commercial chicken farm, overseeing a staff of sixty, and for the first time in two decades with the company, he had been promised a pension. The first thing Al noticed when his father pulled the family's Pontiac station wagon off a barren stretch of Highway 422 and onto the farm's gravel road was the eight aluminum chicken houses. They were a quarter mile long apiece and shimmered in the sun.

The second thing Al noticed was the smell of chicken shit. The place reeked of it, and Al held his nose as he raced through the chicken houses with his brothers and sister. There were 108,000 squawking, pecking chickens and a processing building where more than one million eggs a week from farms in the area were candled and packaged for sale. All six children were put on the payroll, and Al spent weekends collecting eggs and saving the extra cash for college.

After work, he and his brothers would take their guitars down to the processing building because there was an echo in the egg cooler and their folk songs sounded particularly synced. But Al was lonely in Parkman. There was no library or college among the monotonous miles of fields and farms, and at fifteen he decided to find a ride every Monday morning to a well-respected Catholic high school thirty miles away. He stayed with an aunt and uncle during the week, and though he missed his family, the time away from home was worth it to Al. He joined the school's cast of *Bye Bye Birdie* and became president of the local Junior Achievement chapter, where he was taught capitalism but eventually asked his advisers why they also couldn't study socialism.

At home on the weekends, Al worked the farm with a sixteen-year-old Amish boy named Manassas Kuhns, who wore a wide-brimmed hat and a shirt with hooks instead of buttons, in line with the humble Amish dress code. Al was endlessly fascinated by Manassas. Like many Amish boys, he had stopped attending school in the sixth grade, but he could easily fix the pneumatic tube system in the egg processing room.

Manassas believed the world was flat.

"Have you listened to the news?" Al said, exasperated, one Saturday afternoon as they collected eggs. The good ones would be processed and shipped; oversize eggs with double yolks would be turned into liquid or delivered to Al's mother for family meals. "Haven't you heard of our space program?"

It's 1964, Al thought, and the Apollo program promised to land humans on the *moon*.

"It's fake," Manassas said in a heavy German accent.

Al knew that his friend didn't watch television or read a newspaper, but how could a boy who understood the workings of complex machines have trouble believing that scientists were capable of building a spaceship? Al had never met anyone so intelligent with such a limited worldview, and his friend's insular life at a time of great scientific advancement seemed sad and unfair.

"It's true," Al pressed, pushing his cart down the aisle of the chicken house. "You can't imagine that a spacecraft can fly?"

"I don't believe in any of those things," Manassas said, and went on collecting eggs.

The Vietnam War was raging when Al enrolled in Beloit College in Wisconsin in 1969, but the number on his draft card was low and he settled into an academic regimen crammed with classes on government and law. Behind square brown-rimmed glasses, Al was electrified by thoughts of a changing, conflicted country. Robert F. Kennedy and Martin Luther King Jr. had been assassinated, and in Cleveland three years earlier, someone had posted a sign outside of a bar that said, NO WATER FOR NIGGERS, launching a deadly race riot. If the right laws were passed and protected, Al decided, they could level the playing field between the powerful and the weak.

On campus, he quickly made a name for himself. He carried a giant poster of women and children who had been killed in the Vietnam War and pushed for a moratorium on classes after the Ohio National Guard killed four students at Kent State University. He also lobbied to bring beer to the menu at the student union.

Al wasn't thinking much about girls when he walked into the dining hall his freshman year at Beloit, but Mimi Gingold was striking, with brown hair that hung down to her waist, neatly parted in the middle. Al had met her weeks earlier in an American studies class, and now she was standing with a distinguished man in a crisp, three-piece suit.

"This is Al, Dad," she said. "He's from Cleveland."

"Oh." Archie Gingold, visiting for parents' weekend, shook Al's hand. "That's the home of the Richman Brothers."

Al knew about the fine suits made by the famous brothers at their tailoring plant on East Fifty-Fifth Street, but he had never had the money or the need to buy one. He looked at Mimi. "Is your father in merchandising?"

Mimi laughed. "No, my father is a judge."

Al's eyes widened. He had never met a lawyer or a judge. "I'm very interested in municipal government, sir," he said. "Do you mind if I have dinner with you?"

Archie Gingold had spent years on the bench as a juvenile court judge in St. Paul, Minnesota, turning formal adoption hearings into celebrations with cookies and Kool-Aid and keeping delinquent teens out of detention centers by setting up a series of group homes. He would retire in the late 1970s having overseen ten thousand adoptions. "Law," he told the magazine at the University of St. Thomas in Minnesota, his alma mater, "has a benevolent side to it whether we wish to acknowledge it or not."

To Al, everything about Mimi was exotic and new. She was studying to become a teacher and talked passionately about making education "relevant" in inner-city schools. Al had never been out of the country, but Mimi had traveled to Turkey in high school through a student exchange program. Her mother had set up the first hospital candy striper program in St. Paul and served family dinners on linen tablecloths. Mimi had been raised Lutheran, but, like Al, she was starting to question religion. "Do you believe in original sin?" she once asked.

Al was smitten. Just before he proposed during their senior year at Beloit, Al brought Mimi a candy bar called "Powerhouse" and a card that he made himself. There were no lavish words of love and longing, but years later, Mimi would remember Al's perfect line. "Together," he wrote, "we will build a powerhouse."

Al loved to watch his father work, and on the chicken farm, Richard Gerhardstein was a fair boss. He was proud of his position in management, and though he lived on a farm and was constantly checking in at the chicken houses, it seemed more like a grand responsibility than hard labor.

Once, the power went out in Parkman and the farm's electric water pumps stopped working. Chickens started to die, and Al, his father, and the family teamed up with Amish workers to devise a meticulous plan to have the fire department fill up fifty-gallon drums, which sent water flowing directly to the cages. Al stood in the rafters of the chicken house gripping the hose from a fire truck, and when the water came on, he was lifted right off his feet.

Al married Mimi at Beloit in the fall of 1972, traveling to St. Paul for a wedding reception thrown by Mimi's parents, with beef Stroganoff and wine on the menu, and then on to the chicken farm for a party hosted by Al's parents, with pasta and breaded veal. A few months later, Al's mother called. She had always delivered the bad news in the family, but when she phoned just before Christmas, it was the last thing Al was expecting.

"Mom, what's going on?"

"They're closing the farm and we've got to move."

"What?"

"He's worked so hard," Carolyn said quietly.

Al drove home to Parkman. Richard Gerhardstein was always springing from one project to the next, but Al found his father despondent and quickly learned that there would be no severance pay and no pension. After thirty years with the corporation, his sixty-two-year-old father had simply been let go. "They didn't protect me," Richard said, and Al felt sick.

His older brother, James, on leave from the army, drove to Cleveland to confront Richard's boss. Couldn't they find another job back in the dairy in Cleveland? Al's younger sister, Kathy, studying nursing at Ohio University, wrote a letter to the company: Dad has worked a long time, and his manager had assured him of a pension. That didn't happen.

But the company held firm. For years afterward, Al would watch his father lie about his age on job applications to secure low-paying work in a parking garage or an auto-parts store. He was never the same.

Al saw abuse everywhere, in the way his father's company had treated a loyal employee, in the way poor blacks were charged with minor crimes while affluent white-collar criminals got off with a fine, in the way a break-in at the Watergate Hotel seemed to be undermining the office of the president of the United States. Al also believed the admissions test to get into law school was elitist and a touch corrupt, and though he applied to Harvard and Yale, he refused to take the test or pay an admissions fee. He wrote a letter to each school. *Take me on my own terms or don't take me at all.* Most ignored him.

But New York University reached out, eventually offering a full scholarship to study public interest law, and the young husband whom Mimi liked to call a "country bumpkin" moved into a five-story building with a parapet roof in midtown Manhattan. On Sunday mornings, Al and Mimi would splurge on croissants and apricot jam at a Hungarian patisserie near Saint John the Divine, with its Roman arches and rose window that glowed in the sun. But on most days, they gobbled cold lo mein over the kitchen table as Mimi graded papers and Al prepared for class.

He had been assigned to the criminal law clinic at NYU, and to the case of Six Fingers Gibson. According to prosecutors, Six Fingers had picked the pocket of an undercover police officer on a subway platform, stealing one dollar. Charged with robbery, he was being held in jail on Rikers Island in the East River.

Al spent hours crafting a defense. How could a man nicknamed Six Fingers, clearly an expert pickpocket, fail to recognize an undercover officer with a one-dollar bill sticking out of his pocket? Al drew an elaborate diagram of the subway platform to present to the jury in court, and on the day before the criminal trial, he went to the jail to present his legal theory to Six Fingers, an African American man in his forties who had been living on the street.

"You're right," Six Fingers said when Al told him about the plan. "I do pick a lot of pockets, but I didn't pick *that* pocket."

Al sat back in his chair, wowed by his legal prowess and certain he would win his first trial. Six Fingers reached across the table.

"But I do have six fingers," he said, pointing to the nub just beyond his pinkie.

Al was astonished. Earlier, the prosecutor had offered a deal—a guilty plea in exchange for time served. Six Fingers looked at his eager young attorney and apologized. "I know you really want to try this case, and I think you've got a good theory here. But I've got to get home."

Al cut the deal.

The churn in the legal system seemed outrageous to Al, who watched poor defendants cycle through the courthouse again and again for minor infractions, separated from their jobs and families, without access to counseling or rehabilitation programs. The system is rigged, Al thought, and the public has been duped into believing that neighborhoods are safer when jails are full. He could have spent the rest of his life as a public defender, striking quick deals for petty criminals like Six Fingers. But Al wanted to do something more than feed the system.

He wanted to challenge it.

Al started thinking about a job just after his first year of law school, when big-name firms came to campus to court young lawyers with offers of internships. Even classmates in his public interest law program were heading to prominent firms, which promised interesting pro bono opportunities to help clients who couldn't afford a lawyer.

Al decided to interview for a summer clerkship with one of the law firms that regularly recruited at NYU. He dressed in his only sport coat and headed downtown to meet one of the firm's partners, who walked him through the firm's hushed hallways, pointing out the art collection and arresting views of the city. They lunched at a private club, where Al watched lawyers order white wine in the middle of the afternoon.

"Talk big and do big things," the partner said with an air of confidence that Al found intriguing. Maybe he could find his place here, in a high-powered law firm in Manhattan, earning serious money and working his way from a cubicle to an office to a partnership. One of his best friends was already heading into mergers and acquisitions and would one day negotiate billion-dollar oil deals over rounds of expensive vodka.

After lunch, Al roamed the partner's office, nearly convinced that he had found a good fit. On a drafting table in the middle of the room, he lingered over a blueprint of a proposed exit ramp from the Verrazano Bridge to Staten Island, with a miniature house, much like those in the game of Monopoly, fixed on top.

"So what's this?" he asked the partner.

"This was the old lady who wouldn't sell her house."

"What side are you on? Are you representing the old lady or the developer?"

"The developer. Our job is to get rid of her."

Al imagined the woman in her family home, fighting high-powered men in expensive gray suits. "What if she doesn't want to leave? What if she has a connection to the land?"

The partner shrugged. "We're trying to offer a fair number. But, you know, that's progress."

Al decided to be a different kind of lawyer. When summer came, he took an internship at a nonprofit organization that helped ex-offenders, thinking about his father-in-law, the judge, who had once told Al, "One day, people will see the importance of offering everyone a second chance, that people are capable of changing."

When it came time to look for a permanent job, Al reached out to the leading civil rights attorney back in southern Ohio, Robert Laufman, a graduate of the U.S. Naval Academy who had won a landmark case on racially discriminatory lending practices.

Al wrote Laufman a letter, asking for a job. Laufman wrote back, "Civil rights lawyers don't make enough money to hire."

Al and Mimi decided to move to Cincinnati anyway, and with a loan from Mimi's parents, they bought a Victorian house a few miles from downtown for $37,500, throwing her grandparents' oriental rugs across the hardwood floors and planting petunias along the length of the wide front porch. Most of their neighbors were black, and the young couple liked the idea of having children in one of Cincinnati's few integrated neighborhoods.

While his friends settled into jobs in the high-rises of New York City, Al settled into the cramped offices of the Legal Aid Society, where he sued a supermarket chain for discriminating against black employees and landlords for evicting black families. Laufman tracked Al's legal work, and two years later was impressed enough to ask Al to join the civil rights practice, which, in its entirety, consisted of Laufman, a second attorney and Mary Armor, the office manager, who worked in a narrow hallway, squeezed next to a fussy copy machine. Laufman worked on his old dining room table, and Al bought a desk when the YWCA was getting rid of furniture.

He charged clients thirty-five dollars an hour and wore a brown corduroy jacket to work nearly every day, the one he had worn to his job interview in New York City. Once, back in school, when Al rode his bike from NYU to a hearing at a Manhattan courthouse, the judge had looked at Al's scuffed loafers and jacket and called down from the bench, "Are you going camping or are you going to court?"

Al complained about it to his father, but Richard Gerhardstein was unsympathetic. "If you want to be a lawyer you need to dress like a lawyer," he admonished.

A few months after Al joined Laufman's civil rights practice, Mary Armor insisted they go shopping. On a lunch break from a trial, she dragged Al to a downtown department store and helped him pick out a wash-and-wear suit jacket, pants, and a raincoat for $100.

There wasn't much money for salaries, but over six A.M. racquetball games, Al and Laufman talked about their cases against the police, Ohio's prisons, and the City of Cincinnati. Al began to see himself as a legal crusader of sorts in a town that was gaining a reputation as one of the most conservative in the country.

In 1985, Planned Parenthood became a client. As a civil rights attorney, Al believed more than anything in the right to free speech. But as protests at the clinic intensified, picketers would lock arms, forming a hissing, angry wall, often reaching for patients and their escorts as they fought to get inside.

"You can't arrest us because we're praying," one woman told Laufman, whose fifteen-year-old son had been helping women into the clinic.

Al filed a class action lawsuit and listed John Brockhoeft, the man in the ski mask, as the named defendant. Armed with an injunction, Al helped put fifty-seven people in jail for contempt of court, stopping just once when Mimi went into labor with their third child, Jessica. Armor went to court that day on Al's behalf and told Judge Crush, "Al won't be able to come in."

An anti-abortion activist who was sitting in on the hearing said, "I didn't even know that he had children."

"Just because you support Planned Parenthood doesn't mean you're against having babies," Armor said, picturing Al at his desk, scrawling notes on a legal pad with his three-year-old son, Adam, perched on his knees.

Brockhoeft was elusive in the early months of 1987, when the pipe bomb was planted in the bushes outside the Planned Parenthood clinic. But a year later, federal agents found him in northern Florida in a car packed with bomb-making materials. Part of a militant Christian group called the Army of God, he would serve twenty-six months in prison for the thwarted Florida attack and four years for one of the bombings in Ohio.

Al continued to represent Planned Parenthood, even when the archbishop of Cincinnati asked him to stop by for a visit. The Diocese of Providence, Rhode Island, had recently excommunicated the director of Planned Parenthood. "Why are you still representing Planned Parenthood?" the archbishop asked Al. "You're going to go to hell."

"That's not really appropriate," Al replied. "The work of the church in the inner city with poor people and with immigrants gives us plenty to agree about. Let's focus on that instead of reproductive rights."

The archbishop wasn't happy with Al's answer, but on the way out, he gave Al a rosary for his mother. Al thanked him and went back to work.

6

WINNING FOR LOSING

JEFFREY GERHARDSTEIN, Al's brother and the youngest of the six Gerhardstein siblings, fell in love in 1983 with a music teacher named Bob Navis Jr. There was no big to-do about it in the family, no official announcement. Jeffrey told his mother that he was gay, and Carolyn Gerhardstein consulted her uncle Tony, the *medico*, who had once worked as a public health doctor on a South Dakota Indian reservation.

"What is this gay thing?" she asked. "Will it pass?"

"Nothing that Jeff can control," her uncle replied easily. "It was the way he was born."

So Carolyn told Al, and Al told Mimi, and that would have been the end of it, except that the Catholic Diocese of Cleveland found out the two men were living together and fired Bob Navis from his teaching job.

St. Francis Church, in the heart of the city's east side, had withstood six nights of deadly race rioting in 1966, but four years later, a Woolworth's dime store across the street caught fire and the flames engulfed the rafters of the massive stone church. Navis, who had been baptized at St. Francis, rehearsed a production of *Carousel* in soggy rooms covered by sheets of plastic as the church was rebuilt around him.

He had been a music teacher and organist for fifteen years when he met Jeffrey Gerhardstein in a Catholic social group for gay men. Deeply religious, with soft brown eyes and his mother's dark hair, Jeffrey had once considered becoming a priest but decided instead to work in community mental health and help male survivors of sexual trauma. The two men moved in together and set up discreet meetings in another parish with a reverend who had agreed to perform a marriage ceremony. Then Navis lost his job at St. Francis, and in June 1985, his classes at another Catholic high school were also canceled. The Cleveland diocese called it "a matter of church teaching."

Jeffrey didn't know where to turn at first, but he knew his brother would be a good start. Al jumped in, though he hadn't been to church in Cleveland in years and knew very little about the workings of the diocese. "Who's in charge there?" Al demanded. "Can you set up a meeting?"

Though anti-abortion protests were spreading across Cincinnati, Al drove to Cleveland to meet with Sister Christine Vladimiroff, the secretary of education for the diocese, and Reverend John Murphy, the superintendent of schools. It was Al's first gay rights case, and he struggled to find some kind of compelling legal strategy. The law wasn't on his side and he knew it.

Most states in 1985 still criminalized homosexual acts through sodomy laws. The U.S. Supreme Court a year later would find the laws constitutional, with Chief Justice Warren Burger writing, "To hold that the act of homosexual sodomy is somehow protected as a fundamental right would be to cast aside millennia of moral teaching."

Al also knew there was no protection for sexual orientation in federal civil rights law, which banned discrimination based on race, color, religion, gender, and national origin. Three years earlier, Wisconsin had become the first state in the country to prohibit discrimination against gays, with Republican governor Lee S. Dreyfus saying, "It is a fundamental tenet of the Republican Party that government ought not to intrude in the private lives of individuals where no state purpose is served, and there is nothing more private or intimate than who you live with and who you love."

Even if anti-discrimination laws were in place, Al would be arguing before the Catholic Diocese of Cleveland, founded in 1847 and among the largest in the country, with several hundred parishes and schools. He researched canon law but knew even before he began that he couldn't sue the Roman Catholic Church, which could hire and fire its employees at will.

He had to find another way.

On a muggy day in June, Al pulled up in front of a nondescript office building in downtown Cleveland, headquarters for the diocese, and hugged his brother and Navis, who had led rousing singalongs at Gerhardstein family reunions. Navis had been with the family a year earlier when Richard Gerhardstein died of congestive heart failure at seventy-two, ten years after he lost his job on the chicken farm.

"Remember," Al said as they walked into the building, "sometimes change happens when you lose a case you should win."

Al had written a legal brief, but without the backing of civil law, there was no way to argue that Navis had been a victim of discrimination. Instead, Al would focus on a technicality in the teachings of the Catholic Church, and he felt badly for his gentle, perceptive brother, sitting quietly next to the man he loved.

Catholic law stated only that homosexual behavior—not sexual orientation—was sinful. So Al pointed to the two men, who had been planning a wedding celebration for 175 guests. "There is no proof that there is any consummation of their relationship," Al said carefully. "You have no evidence."

The room reminded Al of his boyhood parish in the suburbs of Cleveland. He paused to look at Reverend Murphy and Sister Vladimiroff, who were sitting perfectly still, their expressions blank. Al quickly decided that suggesting Navis and Jeffrey were roommates instead of lovers was likely a losing argument. He moved on.

"Bob has strong ties to the Catholic Church, and his family is devoted. There is no misconduct here, and he is loved by his students."

Two days earlier, 150 supporters had prayed with lit candles on the steps of the Catholic high school where Navis had been a teacher. His parents and Carolyn Gerhardstein were there, along with gay men and women who stepped out publicly for the first time, braving exposure by local television crews. Al was hoping the show of support, unprecedented in conservative Cleveland, would provoke a conversation with church officials.

But Navis lost his job anyway.

Years later, Sister Vladimiroff would refuse to follow an order from the Vatican, which demanded she stop another member of her community from advocating for the ordination of women. But back in 1985, she told the *Cleveland Plain Dealer* that Navis had "publicly espoused a lifestyle that is not in concert with the teachings of the church."

For years afterward, Al would remember Navis's pained face and the words he had whispered just before Al drove home to Cincinnati. "There is no fighting that force," Navis said.

In the winter of 1992, when much of the Northeast was digging out from a massive snowstorm, Al was juggling a heavy load of cases and three children in school. Jessica, his youngest, would learn only later that her father often went back to his law office at night and worked straight through without sleeping to catch up on a backlog of cases. He would come home the next morning for breakfast, and though he was exhausted by bedtime, he would make up elaborate stories about the adventures of her stuffed bear, Boo Boo, gesturing wildly about the bear's ability to save a drowning friend and make the nightly news.

"What were your favorite things today?" Al would ask his daughter. Jessica, with her long brown hair and pale skin, wanted more than anything to look like her father, who had inherited the olive coloring of his Italian family.

In a pink bedroom with bunk beds, a collection of dolls from different countries perched on a shelf by the window, Jessica answered her father the same way every night. "Hugging you and kissing you, having a nice family, and eating dinner."

"Me, too," Al replied in the suit he'd worn to work.

Al kept the details of his cases confined to the office, though he talked to his children about power, why it needed to be poked and tested, and described the powerless people he had come to represent. Just once, years later, Al would cry in front of his oldest son, Ben, when a prisoner he represented was stabbed fifteen times and died behind bars. Ben was moved by his father, who, in the hours after the murder, seemed to be questioning whether he had acted quickly enough, pushed hard enough, to bring safety measures to the unregulated private prison. Al would collect a $1.75 million settlement and distribute it to all two thousand inmates, then push to close the facility.

Al was buried under a long list of cases in the final days of 1992, but he straightened in his chair when Scott Knox called to describe a movement that was brewing in Cincinnati's conservative circles. Knox, a soft-spoken attorney with a handlebar mustache, got his start helping AIDS patients with estate planning after learning that other attorneys in town, worried about the spread of the virus, were covering chairs with sheets of plastic when gay men came in to ask for help. Knox's practice expanded and he got involved in gay rights advocacy, handing out condoms at Cincinnati's gay bars and lobbying for stronger laws.

Now, he told Al, the only law in the city that protected the gay community from discrimination was being threatened by a vocal and organized group of activists. Al thought about Bob Navis, who hadn't gone back to St. Francis since the meeting with the diocese. "We need to do something," Al said, and set up a meeting with Knox.

Over the years, Scott Knox had found many reasons to stay in Cincinnati even as other gay men fled, and one of them was Mrs. Georgia Metz, who was sixty-five years old, spoke in an Appalachian twang, and whittled away the days on her front porch in hair curlers and a flowered muumuu. Knox, who lived across the street in a renovated house with a 180-degree view of downtown, liked to tell Mrs. Metz that she was the most beautiful woman in the world.

Once, a neighborhood college student raced down the street in his jeep, mocking Knox's roommate. "Fag! Fag!" he called into the wind. When the student's friend came by a few days later to ask Mrs. Metz if he could borrow a blender, she promptly replied, "Sure, honey. I'll lend you a blender.

"But those two boys who live across the street from me are the nicest people in the neighborhood, and if your friend wants to call them fags, then he'd better leave, because they're staying."

Knox often thought about Georgia Metz as he counseled men, fired for being gay, who flooded his Cincinnati law office. The attorney with a sign on his office window that read GLBT LEGAL ISSUES started keeping a list of disturbing incidents around town: the hotel that refused to rent space to gay or lesbian groups, the hairdresser who turned away a client with HIV, the gay activist who had been told he would be shot at an upcoming rally.

Back in the 1950s, when Knox was growing up, a former champion swimmer from the University of Cincinnati had decided that pornography was perverting the young and launched an advocacy group, Citizens for Decent Literature. Charles Keating Jr. would one day run a disgraced savings and loan operation that would cost the federal government $3 billion, but in the 1960s and '70s, he spent much of his time trying to rid Cincinnati of offensive movies, theater, books, and magazines. He took a particular interest in homosexuality, which, he later warned, represented a "seduction of the innocent."

In the 1980s, when Knox was finishing law school at the University of Cincinnati, a group called Citizens for Community Values started warning about the pandering of pornography and eventually took aim at the "militant agenda" by homosexual activists. The group was supported by Carl Lindner Jr., a billionaire businessman who was one of the most influential people in Cincinnati.

Knox started practicing law as gay men in Cincinnati were getting arrested in the bathrooms of local parks, where undercover police officers would ask, "What are you into?" and then arrest the men for exposing themselves. But in the early 1990s, Knox started seeing hints of change around town.

It began at Procter & Gamble, one of the city's largest employers. In 1992, the company that sold toothpaste and Metamucil to America's mothers revised its policies to protect gay employees from discrimination. But Knox credited the most remarkable development to the Cincinnati City Council, which proposed a new law that would for the first time protect the gay community from housing and employment discrimination. In a town that had been dubbed "puritanical" by writer John Updike, the proposal was heralded as a major win for the gay community.

More than seven hundred people crammed the hallways of city hall that October to debate the proposed law, known as the Human Rights Ordinance. While opponents chanted, "Jesus, Jesus," a woman in church clothes turned to Knox and said politely, "Don't you understand? I just don't want you to go to hell."

The city council passed the law, expanding the city's anti-discrimination ordinance to include sexual orientation. "This is anti-family, anti-church, anti-God, anti-business," community leader Charles Winburn told the *Cincinnati Enquirer.* "We may have lost the battle today, but we will win the war."

Knox knew the threat wasn't frivolous. In Colorado, voters had approved an amendment to the state's constitution that prohibited the passage of anti-discrimination laws for gays anywhere in the state. Pro-family groups were mounting similar voter initiatives across the country.

Soon after the passage of the Human Rights Ordinance in Cincinnati, conservative groups including Citizens for Community Values launched a campaign to repeal it—and amend the city's charter to permanently ban any new laws protecting the gay community from discrimination. At the helm of Citizens for Community Values was Phil Burress, a former union negotiator who had publicly described his own addiction to pornography and the religious experience that had helped him kick the habit. "If we elevate homosexuality to the same level as heterosexuality by law," Burress later told the *Dayton Daily News*, "then we'd have to do it for everyone—for transvestites, for pedophiles, for adulterers, for rapists, for those who engage in bestiality."

Knox wanted to defend the Human Rights Ordinance, and he needed a certain kind of lawyer to do it. Al Gerhardstein feels civil rights down to the bone, Knox thought, and it doesn't matter that he's a straight guy.

Knox called Al, who immediately agreed to challenge any campaign that would permanently deny legal protections for the gay community. "You can't carve out one section of the population and say, 'You guys, you have no access to the law,'" Al said. "That's wrong and it's unconstitutional."

The push to overturn the Human Rights Ordinance, from powerful corners of the city, was quickly dubbed "Take Back Cincinnati."

Organizers reached out to African American church leaders, who argued on television and radio that gays didn't need protections because they weren't a true minority group, one that was denied the right to vote, sent to segregated schools, or refused access to public facilities. Looking back, Al would find the move to rally black clergy in a city that was 40 percent black deft and unexpected.

Organizers also reached out to lawyer Chris Finney. In 1993, Finney drafted a proposed amendment to Cincinnati's charter using language that was similar to the amendment that had passed in Colorado. During the general election that November, Cincinnati voters would be asked to decide "Issue 3," which not only repealed the Human Rights Ordinance but permanently banned any new anti-discrimination laws for the gay community.

Al was facing a well-funded coalition, and the legal battle ahead of him seemed nothing less than daunting. When it was over, the case would consume nearly five years of his life and, since he worked on contingency, hundreds of thousands of dollars in time and expenses. But over the long, busy summer of 1993, Al was fixed mostly on the image of his children watching television ads sponsored by groups working on the repeal.

One summer night before dinner, Al listened to the conservative commentary on television and told eight-year-old Jessica, "This is just wrong. There's a reason that this group of people needs extra protection."

Jessica wasn't surprised by her father's remark. Her second-grade teacher had been a lesbian and her uncle Jeff was gay. On the bus to school, she had once overheard a boy telling his friend, "Oh man, your pencil is so gay."

From behind them, Jessica shot out, "No. That's not right. Do you mean that kid's pencil is having sex with another pencil?"

"What are you talking about?" the boy said. "That's so stupid."

"Well, that's not the right thing to say."

Now she looked at her father, who was frowning at the television. "There's a history of inequality here," Al said, "and laws against discrimination help stop that injustice."

One day, Jessica would go to law school, prodded by her father. "You know," Al would muse, "you might be frustrated if you're on the sidelines and not able to engage directly in the systems that you want to change."

The campaign to repeal the Human Rights Ordinance raged on in its final months, drawing national media attention. "Homosexuals have made it no secret that pornography and exposing children to homosexual behavior at a very young age are on their agenda," Burress wrote in a Citizens for Community Values newsletter. Supporters of the Human Rights Ordinance pushed back with images of Adolf Hitler and the Ku Klux Klan, upsetting some members of the Jewish community.

Since Knox wasn't a civil rights lawyer, Al teamed up with another local lawyer, Scott Greenwood, who was an expert in constitutional law and a junior associate at a prestigious law firm. When the Human Rights Ordinance passed, Greenwood had decided to take a chance and list his partner's name on his insurance forms at work.

"We have to find a way to kick this off the ballot," he told Al. "It's deceptive what they're trying to do."

Greenwood and Al asked for an emergency injunction, but the

judge ruled against them, clearing the way to put Issue 3 before voters in November.

At Greenwood's townhouse, someone used bolt cutters to sever the electrical connection, and midnight callers whispered, "You're going to rot in hell." The hate mail sent to Al's office reminded him of the years he spent defending Planned Parenthood. But the picketers who had accused Al of helping to kill unborn babies believed they were trying to save lives, and on some level, Al understood and accepted their anger. The callers who were now shouting "Fag!" into Al's voicemail seemed only hateful.

On November 2, 1993, one year after the city council approved the Human Rights Ordinance, voters resoundingly repealed it and permanently amended the city charter, banning any new laws that protected the gay community. Scott Knox took the day off from his law practice to work the polls. "Wow," he thought when the voting results came in. "My city really hates my guts."

7

CASE 773

A BEAGLE mutt named Sparky saw the flames first. They advanced across the roof of Al's house, consuming his son's attic bedroom and sending clouds of charcoal-colored smoke into the cold February sky, just above Sparky's favorite patch of grass and dirt on the front lawn of the house next door. He had been taken in as a puppy by Al and Mimi after he'd wandered into their yard, scrawny and missing a good chunk of his tail. Neighbor Roberta Jackson adopted him, and he spent most mornings sizing up the comings and goings of his quiet, leafy neighborhood, where kids went from house to house shoveling snow every winter.

Just before lunchtime, Sparky started yelping and Jackson stuck her head outside to check on things. The air was thick with smoke, and Al's roof, coated with snow and ice, was burning. Jackson and her late husband, a retired garbage collector, had known Al and Mimi for years. Jackson liked to keep watch over the children when they came home from school and had offered her backyard to Al's expanding vegetable garden. Now she raced to the call the fire department and Al at his law office.

As firefighters prepared to take a hatchet to the front door,

Jackson ran over with a spare key. The house was empty except for the family dog, which was quickly led outside. Mimi arrived home first, and though the fire was out, she gasped at the steaming, gaping hole in the roof. Inside, standing water covered her grandmother's oriental rugs, her mother's linen tablecloths, her children's stuffed animals. The wood floors were buckled. Looks like corduroy, Mimi thought, fighting shock and tears.

Al pulled up next. "Could it have been intentionally set?" he asked one of the firefighters, sizing up their home of nearly twenty-five years. No one was sure.

As arson investigators picked through the rubble, Al thought about the pipe bombs that had been planted at the women's health centers he represented in the 1980s and his latest foray into the volatile world of Cincinnati politics, this time pitted against a powerful coalition of activists that had ushered in Issue 3. After the measure passed, Al and Scott Greenwood had decided to sue the city in federal court.

Neither attorney would be paid for their time unless they won the case and a judge ordered the City of Cincinnati to cover the costs. Al was working entirely on contingency, and every time he signed on new clients, he made sure they knew that reforms—like the monument he would insist go on the front lawn of a police department that failed to protect a young domestic violence victim—were just as important as money. Al had taken loans to support his casework. And now, in February 2004, his house had caught fire.

"Is there any sense that any of those people are involved in this?" Greenwood asked about their opponents on the Issue 3 case.

Al had always shrugged off mild forms of menacing behavior, not wanting to worry Mimi and not particularly concerned that anything would happen anyway. Only once, a year later, would Al ask his family to be cautious, taping John Brockhoeft's picture to the refrigerator when the abortion clinic bomber was released from federal prison. But Brockhoeft was still behind bars when the fire at Al's house started, so he looked at Greenwood and shrugged. "We just don't know."

With a trial over Issue 3 only a few months away, Al settled his family into a neighbor's spare bedrooms and went back to work. Already the case was drawing national interest and a coalition of conservative groups had hired several prominent lawyers to defend Issue 3 in court, including Michael Carvin, a sharp, young litigator who had been a deputy assistant attorney general in the U.S. Department of Justice. One of his cocounsels was Robert Skolrood, with a legal career advocating for Christian fundamentalists as the executive director of the National Legal Foundation, launched in the 1980s by Reverend Pat Robertson.

They would make a formidable team, and over late nights in Al's conference room, Al and Scott Greenwood plotted legal strategy. Though the gay community had celebrated the Human Rights Ordinance, Al had never really thought much about the law because it didn't permit aggrieved parties to file civil lawsuits or require companies to reinstate employees after they had been fired.

But even a flawed law was better than no law at all, and Al had decided it could be sharpened over time. Now that couldn't happen because Issue 3 prohibited the city council from ever passing another law that protected the gay community from discrimination. That prohibition, which applied only to Cincinnati's gay residents—not to women, African Americans, the disabled, the elderly—seemed to Al like prejudice of the worst sort.

He started drafting arguments and over time began to realize that the case, among the first of its kind, would seek to answer the most fundamental questions about what it meant to be gay. He would call on lawyers, political scientists, a psychologist, and gay men and women to explore the ill will that Al believed had targeted Cincinnati's gay community.

He was immensely relieved when the fire department called to say that malfunctioning heating coils, not foul play, had sparked the fire at his house. Just weeks before the biggest trial of Al's career, the house was repaired and Al and his family moved back home.

Your Honor," the clerk in the U.S. district court announced on June 20, 1994, "this is Civil Case 1-93-773."

On the first day of the Issue 3 hearing, Al sat perfectly straight in the quiet federal courtroom, with its wood-paneled walls and dimly lit portraits of retired judges for the Southern District of Ohio. He had been in this room many times before, for cases he could recall with perfect clarity, but this time was different.

He glanced across the room at Michael Carvin, the former deputy assistant attorney general who had traveled from Washington, D.C., to argue the case on behalf of the coalition that called itself "Equal Rights, Not Special Rights." Carvin was sitting with Robert Skolrood, who had pushed to dismantle anti-discrimination laws in Colorado. The room was packed, filled with a mix of African American church leaders, gay rights advocates, journalists, and government officials, including an attorney who would represent the City of Cincinnati. Scott Knox was busy back at his law office, but he would duck in several times over the course of the hearing.

Seventy-four-year-old judge S. Arthur Spiegel, who had fought in the Pacific during World War II and then earned a law degree from Harvard, would hear the case. As a young lawyer in the late 1940s, Spiegel had opened his own law practice in Cincinnati after established law firms turned him down; he believed they didn't hire Jews. Nominated to the bench by President Jimmy Carter, the judge had banned race discrimination in prisons and once allowed himself to be locked in solitary confinement so he could learn what it felt like to be alone in a prison cell. Though some lawyers considered the judge too liberal, Al believed the soft-spoken jurist had an abiding respect for the facts and the law.

"Are we ready to proceed now?" Judge Spiegel asked at nine A.M.

"Yes, Your Honor." Al stood at a table in the front of the courtroom next to Scott Greenwood and two attorneys who had signed on as cocounsels from the national gay rights group Lambda Legal, the oldest and largest gay and lesbian legal organization in the country. Back in Al's office, the desks and floors were strewn with files, last-minute research that would have Al, two interns from New York University, and office manager Mary Armor working late into the night every night of the trial.

Al took a quick breath and gestured to the clients sitting next to him, including several gay residents of Cincinnati. "This case will determine if government can continue to single out these unpopular citizens and deny them basic civil rights. My clients and many who stand with them sincerely hope that the answer from this court will be 'no.' Full rights to free speech and political participation and the right to petition your government must be guaranteed to gay citizens."

"In short, the evidence is going to show you that, one, the factors presented to the Cincinnati voters for consideration in voting on Issue 3 amounted to an appeal to prejudice, an up-and-down vote on whether you like gays. Two, the structure of Issue 3 was designed to cut gay, lesbian, and bisexual citizens out of the political scene in Cincinnati. And three, the reasons posed to justify Issue 3 by the defendants are themselves infected with prejudice."

Michael Carvin stood up. The thirty-eight-year-old lawyer had spent three years in the 1980s as a special assistant in the Civil Rights Division of the Department of Justice. But in private practice his libertarian leanings prevailed, and over the course of his career he would sue the federal government on a number of fronts.

Carvin believed it had been infinitely rational to give the citizens of Cincinnati the same authority as legislators to decide the law on gay rights—an argument for the democratic process that would be heard again, years later, when Jim Obergefell and John Arthur filed suit against the State of Ohio.

Early in Carvin's career, he had learned not to personalize litigation and had honed an amiable but relentless approach in the courtroom. Now he turned to Judge Spiegel and said easily, "The plaintiffs have emphasized this morning, as they have emphasized throughout this case, they have claimed that they've lost some fundamental right to political participation."

"Now remember," the judge interjected, "we aren't arguing—"

"I understand, Your Honor."

"I know it's not easy to do."

" . . . What the evidence will show in this case, Your Honor, is that Issue 3 was not discrimination. Issue 3 was not invidious, and homosexuals are not a group entitled to the special protections of the Constitution."

Al, who would present his arguments first, called John Burlew to the witness stand. The African American trial lawyer was a member of the Ohio Civil Rights Commission and had followed the Issue 3 campaign closely. He would later become a municipal court judge.

Burlew looked at Al and described the campaign. "The message that . . . you saw appealed to the base worst emotions, I think, in people. The public debate that I was involved in primarily would result in a spokesperson relying on a passage of Leviticus or something talking about homosexuality is an abomination. That's the word I'll never forget coming out of that debate."

After lunch, Carvin responded in rapid fire. "So not everyone who supported Issue 3 was irrational or a bigot, isn't that correct?" he asked Burlew.

"I don't want to even begin to suggest that. Of course not. A lot of honorable people supported it."

"Right. You're not saying you have a monopoly on all wisdom in Cincinnati, are you?"

"I would state the exact reverse of it. I don't consider myself in that posture at all."

"You think reasonable people could disagree on this issue, don't you?"

"Absolutely."

Carvin pressed on minutes later. "And you yourself are tolerant of other beliefs and lifestyle[s], are you not?"

"I hope I am."

"You don't agree, for example, that homosexual behavior is an abomination, do you?"

"No, I don't."

" . . . Nothing wrong with expressing religious viewpoints in public, is there?"

"Absolutely not."

"It's part of a long and noble tradition, including Martin Luther King, isn't it?"

"Absolutely."

"And you don't want to restrict the religious freedom of these people?"

"Absolutely not."

"You don't want to force them to keep their religious beliefs to themselves, do you?"

"No."

"Because religious freedom includes more than having beliefs in your head. It includes the ability to act on those beliefs through political participation and that sort of thing, doesn't it?" Carvin asked.

"Correct," replied the African American lawyer. "It also justifies slavery and other things which I disagree, all of which have biblical and religious basis."

On the second day of the hearing, Al wanted to show Judge Spiegel that Issue 3 had been driven by distaste and even hatred for the gay community, so he called Kenneth Sherrill, a political science professor at Hunter College in New York and an expert on gay and lesbian politics. "Sir, are you saying that people just don't like gays?" Al asked.

The professor described how a number of political scientists had been measuring public opinion since 1964. Gays, he said, were among the most disliked and politically powerless groups in the country. "There is no group in American politics that has, over such a long period of time, gotten systematic responses of cold feelings or feelings of emotional distance."

On the third day, Al questioned African American lawyer Jerome Culp, the son of a Pennsylvania coal miner who had earned a Harvard law degree and gone on to teach employment and labor law at Duke University. Al set up an easel near the judge with the Issue 3 wording displayed on poster board.

> The City Council may not enact, adopt, enforce or administer any ordinance, regulation, rule or policy which provides that homosexual, lesbian or bisexual orientation status, conduct or relationship constitutes, entitles or otherwise provides a person with the basis to have any claim of minority or protected status.

"Is Issue 3 a typical government civil rights provision?" Al asked the professor.

"No. . . . What that really does," the law professor said, "is to single out homosexuals, lesbians, and bisexuals, in some sense to sort of put a scarlet H on their forehead, and to say that it's all right to both individually and institutionally discriminate against them."

"Based on your study of the history of discrimination and civil rights," Al asked seconds later, "do you have any knowledge as to whether there's a common prejudice against gays, lesbians, and bisexuals?"

"There's a lot of prejudice against gays, lesbians, and bisexuals."

"Objection," Cincinnati's attorney, Karl Kadon, called.

The judge responded dryly, "The Court will take notice of the fact that he's sitting in this community and there is such things. Overruled."

Later, Michael Carvin stood up. "Let's substitute the words *child molesters*, okay, for *homosexuals, lesbians or bisexuals*. Issue 3 is exactly the same except I'm saying . . . *child molesters*. City council can't pass laws having quotas for child molesters or protected status for child molesters. That cuts off the city council's future options in terms of protecting child molesters, does it not?"

The law professor, knowing it was highly unlikely that any government in the country would create anti-discrimination laws for child molesters, said, "Mr. Carvin, you live in a world that I don't."

That afternoon, Carvin called one of his own witnesses, the Cincinnati lawyer who had drafted the ballot language for Issue 3. Al would always remember the look on Scott Greenwood's face when Chris Finney took the stand and said, "Let's say there's a company picnic and someone wants to go to the company picnic and have a wholesome atmosphere where they can bring their family and expose them to the kind of moral atmosphere that they display in their home life. And yet, because of the city ordinance, we now have to have homosexual couples come to that company picnic exposed to our children and not feel as comfortable working in that atmosphere."

Al had taken on the case as a civil rights lawyer who believed that gay people deserved to be fully engaged in the community. But Greenwood was both a lawyer and a gay man, and now his very identity was under assault. Just before the trial started, Greenwood had lost his junior associate position at the law firm where he worked. The partners offered no particular reason for it, but Greenwood assumed his work on Issue 3 had made some clients uncomfortable.

When Greenwood stood up to cross-examine Chris Finney, Al briefly wondered whether his cocounsel's head was clear. "Can you tell me, Mr. Finney," Greenwood asked carefully, "how sexual behavior is somehow related to whether a gay, lesbian, or bisexual person can eat in a restaurant or work in a job?"

The most contentious day of the trial came on the fifth day, when downtown buildings were getting ready to set out flags for the Fourth of July. Michael Carvin called Clemson University political science professor James David Woodard, who described gays and lesbians as "quite an active and successful and powerful political group."

Al was agitated. "You'd agree that gays have experienced a history in this country that includes deliberate exclusion from certain occupations like the military, sensitive government posts, and various occupations, right?" Al asked the professor.

"Yes, sir."

"And you'll agree that for many years, homosexuality was viewed even as a mental illness?"

"Yes, I believe it was."

"And you'll agree that there was a period and continues to be allegations of harassment by law enforcement agencies against gays, right?"

" . . . Allegations, yes."

"Of additional harassment by law enforcement agencies; is that correct?"

"Objection," Carvin called. "It's a compound question."

"We'll do it in two pieces," Al said. "In the past, the police have harassed gays, right?"

"As much as I know, yes."

"Okay. And there's a continuing concern about fairness toward gays on behalf of law enforcement agencies, right?"

"I guess."

Later, Al switched gears. "Should a heterosexual who is engaged in oral sex be denied the right to vote?"

"No," the professor said.

"And there's no history of that type of restriction in this nation, is there?"

"Not that I know of, no."

"Should a homosexual who is engaged in oral sex be denied the right to vote?"

"No one should be denied the right to vote, Counselor."

"Should a heterosexual who has engaged in oral sex be denied the right to hold public office?"

"No."

"And should a homosexual who has engaged in oral sex be denied the right to hold public office?"

"Don't know of anyone who is denied. No."

Al pushed forward. "Should a heterosexual who is engaged in oral sex be permitted to petition the Cincinnati City Council and seek civil rights legislation?"

"Lobby city council? I suppose so."

Al paused. "And should a homosexual who has engaged in oral sex be permitted to petition city council and seek civil rights legislation?"

"Sure."

Al had what he wanted. "Now, would you agree that after Issue 3, city council has jurisdiction to act on the petitions of the heterosexuals, but not on the petitions of the homosexuals for civil rights legislation?"

Michael Carvin rose quickly. "Objection. Mischaracterizes."

"Overruled," the judge said.

From the witness stand, the professor said, "This is getting a little bit out of my league."

Six weeks later, Judge Spiegel struck down Issue 3 in a seventy-five-page decision, calling the ban on anti-discrimination laws for gays unconstitutional. "Despite the fact that a majority of voters may support a given law, rights protected by the Constitution can never be subordinated to the vote of the majority. While at times this may seem unfair, especially when deeply emotional issues are involved, indeed it is the fairest and most deeply rooted of all this nation's rich traditions."

The backers of Issue 3 quickly announced they would appeal.

"The big issue here is who is running this country," Phil Burress told the *Cincinnati Enquirer*. "The government has gotten too big for its britches."

8

AGONY OF LAW

STEPHEN OLDEN first met Al Gerhardstein in a dump of a law office, where roaches scampered up the walls and across the metal desks and a wheezing air conditioner could never quite cut the humidity on August afternoons. Olden had graduated from Boston College Law School and was practicing consumer law at Cincinnati's Legal Aid Society. He worked next to Al in a converted public-housing apartment in the city's west end, counseling indigent families whose stories over time became a sorrowful blur of abuse and desperation. Olden stayed long after Al moved on, but the two young lawyers swapped legal strategy on Saturday mornings, hunched over Al's vegetable garden picking tomatoes, zucchini, and green peppers.

When winter came in 1994, Olden didn't hear much from his friend. Al was preparing to defend Judge Spiegel's decision before the Sixth Circuit Court of Appeals, considered by some one of the most conservative federal courts in the country. A win would help shape national precedent, delivering a major victory to the gay rights movement at a time when other cities were considering similar laws that banned protections for gays. But Olden was worried about his hard-charging friend, who was working long hours without pay and seemed to consider himself singularly responsible for the well-being of Cincinnati's beleaguered gay community.

From the very beginning, Olden knew that Al was driven by a visceral sense of right and wrong, an instinct so deeply rooted in his gut that there was never much room for doubt or second thought. Once, years earlier, Olden had slipped into the back row of a federal courtroom to catch the final moments of a drug product liability trial that had brought big-name attorneys to Cincinnati. He decided to leave early to walk back to the office, but in the hallway outside the courtroom, a federal marshal placed a firm hand on Olden's shoulder and said, "I'm under orders to detain you."

Olden flinched. "You've got to be kidding."

"You left the courtroom during closing arguments."

"But there's nothing criminal about that," Olden insisted as he was pushed toward a holding cell for prisoners in the upper level of the federal courthouse.

The room, with steel bars and a narrow metal bench, smelled of disinfectant. Olden handed over his wallet and briefcase, eyeing the telephone on the wall and briefly thinking about the dozens of inmates he had represented in crowded, crumbling prisons. "Can I at least call a lawyer?"

Al picked up the line. "I'll be right there," he told Olden, then yelled over his shoulder, "Holy shit!"

The detainment order had come from Judge Carl B. Rubin, who had been nominated to the federal bench in 1971 by President Richard Nixon. He was considered fair but unflinching, setting strict rules for the lawyers who argued in his courtroom. Olden had been in Judge Rubin's court before and gasped when the judge's law clerk showed up in the holding room and said, "Just so you know, you've been charged with contempt."

At the beginning of the day, the clerk explained, Judge Rubin had warned spectators at the liability trial not to leave the courtroom while the attorneys were delivering closing arguments. Olden had come in later and didn't get the warning. The clerk just shrugged.

During the lunch break, Judge Rubin had decided to weigh the matter in the building's largest and most ornate courtroom, typically used by judges on the Sixth Circuit Court of Appeals. Al showed up just as the impromptu hearing was about to begin, breathing hard from his sprint to the courthouse. Olden wasn't sure what to expect from Al, who argued most of his civil rights cases in federal court and likely wouldn't be too keen on aggravating a veteran judge.

Judge Rubin glowered from the bench and announced that Olden had violated the rules, leaving the courtroom while attorneys were speaking.

"With all due respect, Your Honor," Olden replied. "I wasn't aware of your rule."

The judge frowned.

Al stood up and said sternly, "Your Honor, this hearing is totally unnecessary. Mr. Olden did nothing wrong."

Al suggested that the judge could have simply pulled Olden aside and asked for an explanation rather than detaining him in a holding cell and threatening contempt charges. The judge relented and asked Olden to write a letter of apology.

Over burgers at a grill just down the street from the courthouse, Al asked casually, "Hey, do you think we can sue this guy?"

Olden was still reeling from the morning's events and at first wasn't sure whether he understood Al's question.

"If he acted outside his jurisdiction, there may be a cause of action against him," Al continued.

Olden knew that federal judges had immunity when they were

acting in a judicial capacity on matters within their jurisdiction. He also knew that Judge Rubin was highly respected and that Al's cases hinged on decisions in federal court. "Probably no case here," Olden said, shocked that his friend would even consider suing.

Al's bold question stuck with Olden for years, and he wasn't the least bit surprised when he learned that Al had taken on the powerful coalition that had successfully banned all laws meant to protect the gay community from discrimination.

Olden often thought about Al and his big case before the Sixth Circuit throughout that long winter, through Thanksgiving and Christmas and New Year's, when it seemed all of Cincinnati was waiting to find out how an influential panel of judges would define what it was like to be gay.

Some of the most successful civil rights cases have started with a story.

One came from Topeka, Kansas, in the 1950s, when a black railroad welder couldn't understand why his daughter had to ride the bus to a segregated school when a perfectly good one, filled with white students, was only blocks away. Another came from San Francisco in the 1970s, when Chinese American parents told the courts that their children were languishing in public schools, unable to learn because they didn't speak English.

Before Judge Spiegel in district court, Al had told the stories of gay plaintiffs like Roger Asterino, a forty-three-year-old city worker who had been mocked by a colleague at work. But arguing in an appellate court was different. There would be no witnesses to query, no easy way to bring stories to life. Attorneys were there mainly to answer questions posed by the judges, and Al knew that the best appellate lawyers listened carefully, answered succinctly, and guided the conversation back to their most convincing arguments.

There were more than twenty judges on the Sixth Circuit, one of thirteen circuits in the United States courts of appeals. Most were assigned to large geographic regions and all had substantial influence on federal law. The Sixth Circuit decided cases from Ohio, Michigan, Tennessee, and Kentucky in the same federal courthouse where Al had defended his friend Steve Olden.

Cases were randomly assigned to three judges, and for weeks Al waited to find out whether the panel that would oversee Issue 3 bent right or left, though he knew that political leanings were never an absolute indication of how a case would play out.

Luck of the draw, Al thought grimly, when he learned ten days before the March hearing that two of the judges assigned to the case had been nominated to the bench by President Ronald Reagan. The third judge, Cornelia Kennedy, had been nominated by President Jimmy Carter despite criticism from left-leaning groups. Twice, Reagan had considered Judge Kennedy for a seat on the U.S. Supreme Court.

Nine months after Judge Spiegel declared Issue 3 unconstitutional, the three judges called the hearing to order.

Al had always loved the grandness of the Sixth Circuit courtroom, with its red silk drapery and walnut-colored walls carved like lattices. He sat next to Scott Greenwood and their two co-counsels from Lambda Legal, just across the way from Michael Carvin, who would speak first. "What they're saying here is: It's wrong for the people to make a perfectly constitutional decision," Carvin told the judges.

Al countered, "This case is about the right of each citizen to full and equal participation in the political process."

Al talked about the hearing before Judge Spiegel, who had decided that gays were an identifiable class, often maltreated and in need of legal protection. Minutes into his comments, Judge Kennedy, the second female judge to sit on the Sixth Circuit, cut in.

"It's pretty hard to identify in any individual, is it not?" she asked about homosexuality. " . . . How are we even going to identify who they are?"

The question surprised Al, who paused briefly. "Your Honor, for what purpose?"

"Well, for any purpose," Judge Kennedy said. " . . . It isn't something that you can look at and determine when you first look at people. . . . We have an identifiable class here that would be almost impossible to identify. We have estimates that they range from 3 percent to 15 percent [of the population], which indicates an inability to determine who belongs to the class."

Years would pass and Al would remember the impulse he had at that precise moment, an overwhelming need to turn to the people around him in the courtroom and ask, "Will everyone who is gay please stand up?" He knew there would be a cross-section of the most ordinary-looking people, including Scott Greenwood, with his leather briefcase and law degree.

Al replied, "The record is real clear that whether you can identify a gay person on an individual basis or not, gay people in this country are the group that a lot of people don't like. . . . There's a widespread prejudice of gays, and that has expressed itself even to the point of excessive violence."

The hearing was unsettling, and Al wasn't surprised when the Sixth Circuit, on May 12, 1995, overturned Judge Spiegel's decision by unanimous vote, becoming the highest court in the country to validate the exclusion of gays from anti-discrimination laws. "The reality remains that no law can successfully be drafted that is calculated to burden or penalize, or to benefit or protect, an unidentifiable group or class of individuals whose identity is defined by subjective and unapparent characteristics, such as innate desires, drives and thoughts," the judges wrote.

They also noted that Issue 3 furthered a "litany of valid community interests," including the removal of possible sanctions against anyone "who elected to disassociate themselves from homosexuals."

Michael Carvin told the *New York Times*, "In my view, this is a victory for democracy, a victory for the right of local communities to decide this controversial issue for themselves."

Al had lost plenty of cases, but the language in the Sixth Circuit's decision seemed unusually harsh and unsympathetic. "The opinion is miserable," he wrote to his cocounsels. "It made me sick to read it."

Judge Spiegel had ordered the City of Cincinnati to reimburse Al and his three cocounsels a total of $375,000 to cover their time and expenses, but the judges on the Sixth Circuit panel overturned that order, too, which meant that if Al and his legal team wanted to appeal the ruling, they would have to continue working without pay. From his office at the Cincinnati Legal Aid Society, Steve Olden called his friend. "The Supreme Court is going to turn this around," Olden promised.

From the beginning, there had never been any question that Al would take the case to the U.S. Supreme Court, and all that summer, he started drafting and redrafting a petition to appeal. He had ninety days to craft nine thousand words that would convince the highest court in the land to hear the case. The Supreme Court at the time received about 6,500 appeals a year, but accepted only about ninety cases for oral argument. If the court rejected Al's petition, there would be no further recourse. Issue 3 would stand.

In August, shortly before the Supreme Court deadline, Al heard from the attorneys at Lambda Legal. They had worked together for months, but now the national organization wanted to take the lead in the case. Al was stunned.

"We do not agree to this change," he wrote. "... Our Cincinnati clients—the folks that have to live with Issue 3—continue to be reassured that Cincinnati counsel are coordinating their presentation."

Lambda Legal attorney Patricia Logue replied, "There are ... much greater stakes for lesbians and gay men generally that are not limited to the impact on our Cincinnati clients."

Al and Scott Greenwood had decided early on to bring in national experts from Lambda, which had spent years advocating for the rights of the gay community. But Al always believed he would manage and lead the lawsuit. It was new and sticky terrain, a local lawyer with a national case and a skilled advocacy group that wanted to make sure it was done right. The legal director for Lambda Legal relented a few days later.

"I want to win our case," Al wrote. "I know you do, too."

He would remember the brief skirmish when he took on another gay rights case from Cincinnati that seemed poised to reshape federal constitutional law, this one filed in the name of Jim Obergefell and his dying husband, John.

Al and Mimi had always liked the Cincinnati Nature Center, on the edge of a deciduous forest with monarch butterflies and a pond full of turtles. Cherry trees bloomed in the spring, and every fall, the leaves on the oak and maple trees turned fiery red. In May 1996, after three years of work on the Issue 3 case, Al took the day off to spend time with Mimi, hiking along miles of winding trails. They had been married for more than twenty years, living at times off Mimi's meager teaching salary so Al could pursue civil rights cases entirely on contingency.

In the car on the way home, there was news about the Supreme Court. Al tensed and turned up the radio, glancing quickly at Mimi, who more than once had finished family vacations alone when Al left to tend to legal emergencies. He didn't want work to creep into their final few moments together, before three kids, the family dog, dinner, and homework shattered the easy silence. But two months earlier, the Supreme Court had "held" the Issue 3 appeal because a similar appeal had come out of Colorado, where all branches of government were prohibited from passing anti-discrimination laws for gays.

Al jumped in his seat when National Public Radio announced that the Supreme Court had struck down Colorado's law. Justice Anthony Kennedy, raised Irish Catholic and nominated to the court by President Reagan, broke ranks to side with the more liberal judges, writing in a majority opinion that the ban in Colorado lacked any legitimate governmental purpose. "It is not within our constitutional tradition to enact laws of this sort," he wrote.

It was the single biggest victory in the history of the gay rights movement to date and seemed to hint at the beginnings of a shift on the Supreme Court, which had found only a decade earlier that laws criminalizing sodomy were constitutional. The decision in Colorado would immediately undercut similar initiatives across the country, including Issue 3. "This is really just totally exciting," he said to Mimi, fighting the urge to call Scott Greenwood and his cocounsels.

Mimi nodded and smiled. She had been following the case since the beginning and wanted Al to win, but looking at her distracted husband after a long afternoon away, she thought, Does this always have to happen when we're having a good time?

Less than a month later, without calling for a hearing, the Supreme Court sent the Issue 3 case back to the Sixth Circuit for reconsideration. "You've got a second shot," Steve Olden told Al that fall. "They can't turn you down now. The Supreme Court is on your side."

Al flew to Lambda Legal headquarters in New York to rehearse what he would say during his second appearance before the Sixth Circuit. This time, he would set up two three-foot posters—one showing the wording of the Colorado amendment, which had been declared unconstitutional, the other of Cincinnati's Issue 3. Al highlighted and underlined the identical words and phrases.

When he walked into the courtroom in March 1997, Al was feeling bold and hopeful for the first time in months. Michael Carvin could sense the change in Al's demeanor. "Ninety-nine percent of the country," Carvin would say later, "would have confidence" after the Supreme Court's decision on Colorado.

But the same panel of judges overturned Judge Spiegel's ruling for a second time, delivering an entirely unexpected win to the backers of Issue 3. Al and his legal team immediately appealed to the Supreme Court, rushing copies of the petition to the courthouse in Washington, D.C. Weeks passed with no word from the court, lousy, meandering days spent waiting in a law office stacked floor to ceiling with research on homosexuality. Finally, Al received a letter from the Supreme Court.

The justices declined to take the case.

The decision was so crushing, so final, that Al could barely catch his breath. As he drove around town to all the familiar places, to the courthouse, to the nature center, to schools and shops, Cincinnati seemed like an oddly foreign place, an enclave of intolerance that had allowed a backward measure to exist and thrive, embedded in the city's charter.

Civil rights cases were never easy wins, but Al felt as if his faith in the law had been shaken to its core, and he wondered whether he could keep on. The country's legal system, predicated on precedent, didn't seem to apply to the gay community. Al was forty-seven and had spent nearly five years working on the case without pay. He thought about packing up and leaving Cincinnati, maybe becoming a law professor in Mimi's hometown of St. Paul. Al would be closer to his father-in-law, Judge Gingold, who kept a neat stack of Al's legal briefs in a cardboard box.

"Justice just didn't happen here," Scott Knox told Al. "You did everything you could."

The dinner table in the Gerhardstein household was usually a happy place, with one kid awarded the "you are very special" plate for a birthday, or a celebration for a win on the baseball field or an extra A on a report card. But Al was sullen and distant, and Jessica, in seventh grade, knew without asking that something was terribly wrong with her father. She looked at Mimi, who said carefully, "Daddy doesn't always win."

It seemed to Al as if there was nothing to celebrate, but a few months after the decision, he got an invitation from his church. For years, Al and Mimi had been taking their children to the First Unitarian Church of Cincinnati, in a historic stone chapel with wooden pews that flashed blue, gold, and pink when the sun settled behind the stained-glass windows. Many of Cincinnati's more liberal thinkers showed up for services, including Steve Olden and his family and Al's longtime law partner, Robert Laufman.

Olden and public relations executive Linnea Lose, who had joined the church as its first openly gay congregant twenty years earlier, had decided to write to the Unitarian Universalist Association in Boston, describing Al and his work on Issue 3. The president of the association wrote back, and over an annual congregant lunch in the church's multipurpose room, his letter was read out loud. "Like most work of dismantling oppression, there are roadblocks and setbacks. The task of true faith-based anti-oppression work is to keep struggling for justice and to persevere even in adverse times. You and your work are testimonies to this process."

Members of the church started applauding, and Al was called up to the podium and asked to speak. It was the first time he was publicly recognized for his work on the case, and Mimi choked back tears as she looked at her husband, who, no matter what anyone said, considered himself singularly responsible for the defeat in court. Finally, she thought, validation after years of grueling, unpaid legal work.

Al shifted from foot to foot, looking at Mimi and the members of his church, who were cheering and shouting his name. For a few seconds, he just stood there, weeping. "Boy," he finally said, with a brief, grateful smile, "I never got an award for losing anything in my life."

9

"MINDS HAVE CHANGED"

THE BEAT cop with eighteen years on the police force had taken to wearing pink lipstick and nail polish, and now he sat in Al Gerhardstein's law office, wondering whether his sudden demotion was grounds for a federal discrimination lawsuit. Philecia Barnes had still been Phillip Barnes in the summer of 1999 when he failed the probationary period for new sergeants at the Cincinnati Police Department, despite a history of good reviews.

"You are being mistreated," Al told the soft-spoken officer, who was transgender and would seek counseling for sexual identity issues. "We'll do the best we can to come up with an argument to represent you, but don't be surprised if we're not successful."

It had been nearly two years since the Sixth Circuit's ruling on Issue 3, and Al was still steaming over a decision that he believed had been fueled more by intolerance than law. He had searched for some kind of legal workaround long after the loss in court, but Cincinnati was now the only city in America with a charter that expressly prohibited anti-discrimination laws on the basis of sexual orientation. Al looked at his new client, who had a degree in social work, and said, "The law is terrible."

But Al pushed forward anyway. In February 2003, after Phillip Barnes transitioned to a woman and changed her name to Philecia, Al and his forty-one-year-old law partner, Jennifer Branch, took the case to trial. There were forty potential jurors, mostly white and mostly from rural, southern Ohio communities. Looking at their faces, Al thought about the lawyer friends who had told him again and again that there would be no sympathy for an African American transgender police officer in the city of Cincinnati.

As Al sized up the potential jurors, he remembered Judge Cornelia Kennedy's unexpected questions about homosexuality during the Issue 3 hearing in the Sixth Circuit. This time, Al would attack the issue head-on. He turned to the jury pool and said, "This case raises sensitive issues about gender, sexual orientation, and perception. . . . Rate yourself on a scale of one to five, one being accepting and five being unaccepting."

Only a handful of people said five.

Surprised, Al continued, "If you had a choice of where you would work and live, assuming it's all safe, and one choice was a community of people that shared your religion, lifestyle, and values and the other choice had people who did not share your religion, lifestyle, and values, would you want to live in the same community or a different community?"

Most people in the jury pool responded, "Different."

The trial lasted for eight days. On the last day, Al stood before jurors and said, "This case is really about holding government accountable. We give government a lot of power in this country. It's only through trials like this that citizens sitting as jurors can make sure that government itself follows the law."

The jury deliberated for six hours before siding with Barnes, awarding her $320,000 in compensation and reinstatement as a sergeant. The decision energized Al, and when he heard about a small group of clergy and activists who were quietly exploring ways to repeal Issue 3, he decided to get involved for a second time.

He went to a meeting to describe the way jurors had responded to Philecia Barnes. "We think minds have changed," Al told the group.

A ragtag group of organizers had started meeting in secret in those early months of 2003, wary that word of a repeal effort would instantly rally the coalition that had fueled Issue 3. The most prominent group, Citizens for Community Values, had recently reached out to Cincinnati's schools superintendent about the "legal liability associated with homosexuality education in public schools."

"It is our hope that you, who have been entrusted with the education and care of children, will carefully consider the numerous negative physical, mental and emotional consequences directly related to homosexual behavior," Phil Burress wrote in a two-page letter.

Those first few meetings about the Issue 3 repeal effort seemed more like prayer sessions than strategy sessions to attorney Scott Knox. Overturning the law would require hundreds of thousands of dollars and an unprecedented call to action by the gay community and its supporters. They would have to kill the law the same way it had passed a decade earlier, through a voter initiative that changed the city's charter. It was an ambitious plan with bad timing: Massachusetts was on the verge of becoming the first state in the country to allow gay couples to marry, and conservative groups were pushing voters in a series of states, including Ohio, to pass constitutional amendments that banned same-sex marriage.

In a local reverend's living room, Knox huddled with a real estate developer, a rabbi, a philanthropist, and a demographer from Procter & Gamble, which worried that Cincinnati's conservative reputation was scaring away top scientific talent. There was also a twenty-four-year-old Catholic man, estranged from his parents, who had discovered in college how it felt to drive change.

Chris Seelbach had grown up an only child in a middle class, postwar neighborhood in Louisville, the fair-haired son of a construction equipment supplier who coached the local baseball and soccer teams. The day after Seelbach graduated from Catholic high school, his mother sat him down in the family's great room and said, "Please tell us you're not gay." Seelbach fled the house, past a two-and-a-half-foot wooden cross hanging in the entryway. A few hours later, he met his mother in the parking lot of a local church and said out loud for the first time, "I'm gay."

Seelbach called a crisis hotline and spent his last summer at home in Christian counseling, sitting through "conversion therapy" for homosexuals. He left for Xavier University in Cincinnati lonely, exhausted, and worried about his parents, whose silence that summer had turned a happy house into a place of strangers. *How could I do this to them? They've done everything right.*

In college, Seelbach and his parents rarely spoke, and though they sent money for tuition, Seelbach took odd jobs to help support himself. But he found his voice on campus, convincing the student body to sign on to an anti-discrimination policy for gay students and teachers and helping to create a gay-straight alliance at the Catholic university.

Seelbach was in law school when he heard about the Issue 3 repeal effort and decided to attend a meeting in the basement of a Presbyterian church. A plan emerged. While Al drafted the repeal language that would be put before voters during the general election in November 2004, organizers would rally support from bank presidents, corporate executives, and the Archdiocese of Cincinnati. Seelbach and hundreds of volunteers would go door to door. "You have to say the word *gay*," organizers instructed during hour-long training sessions. "Practice saying the word *gay*."

Seelbach canvased the city, asking, "Do you think someone should be fired because they are gay or lesbian? Do you think someone should be denied a place to live?" Over time, he knocked on five thousand doors.

The *Cincinnati Enquirer* would later call the effort "the most sophisticated grassroots political campaign in the city's history."

Right up until midnight on November 2, until the television newscast announced the results and the bar packed with campaign volunteers erupted in cheers, Seelbach was still preparing for a loss.

But Cincinnati voters overturned Issue 3, a decade after it had become law.

Outside, crowds of supporters started cheering in the streets. Inside, Seelbach hugged strangers and friends, proud of his city, proud of himself. We actually won something, he thought. Scott Knox, who had stopped counting how many times he stood on front porches saying the word *gay* to grandmothers and religious folk, shouted into the crowd, "This really worked!"

Two nights later at a celebration at the Unitarian church, hundreds of people lit candles. Every seat was taken and dozens squeezed shoulder to shoulder in the aisles and lobby, just beyond a long Tiffany window. Reverend Sharon Dittmar, who had a master's degree from Harvard Divinity School, stood in front of the room and said, "Rest in the victory that many of you never thought you would see. Celebrate this victory."

But that same election, when Cincinnati voters overturned Issue 3, the voters of Ohio had passed a constitutional amendment that banned same-sex couples from marrying. Voters in all eleven states with anti-marriage campaigns had approved the bans, a painful blow to the gay rights movement.

Al wanted to celebrate the defeat of Issue 3, one of the most divisive laws in the history of Cincinnati, but his thoughts drifted to Ohio's new marriage ban and the 3.3 million people who had supported it. The law not only prevented gay people from marrying, but also from having their marriages recognized even when they had been legally performed in other states. Al wasn't entirely sure what that meant or how it would affect gay couples, which is why, looking back, Reverend Dittmar's words at the church celebration that chilly November evening seemed particularly telling.

"The arc of the universe," she said, "must be at least long enough to include a party, lots of sleep, an unredeemable novel, even several of them, a vacation, and staring at the ceiling for no reason—before we begin again."

10

DIAGNOSIS

TO JOHN ARTHUR, the worst part of nearly dying over Mother's Day weekend in 1995 was the embarrassment of it. The asthma attack had left him feeling feeble and exposed, like the beanpole of a boy who hid in a tree when he couldn't finish a physical fitness test at school. Years later, he could still hear the laughter as he darted off the running track, louder with the passage of time and the embellishment of memory, and he never wanted to feel that way again.

Once, while driving through town with his aunt Paulette, John had confided, "I always hated phys ed. I was the target of dodgeballs."

Paulette shot back, "But look what you've become. You're better than any of them."

And she meant it. John wasn't in love with his consulting job in technology, but his job had fueled his passions. He had grown into something of a local tastemaker and filled his life with whimsical, interesting things, fabric and furniture, clothing and paintings. Saturdays with John were grand adventures, a long brunch, a lazy walk, a hunt through the dusty aisles of an antique shop, where John, from the corners of his mouth, would issue a running, witty commentary on his interesting finds and the god-awful items that he rejected.

Lonely people moved him, perhaps because they reminded John of how different he felt as a boy, and more than once he touched his friends by inviting strangers sitting alone at bars to join them for drinks or dinner. Year after year, John brought new people into his social fold. He never shared particularly personal stories about himself, but he was a perceptive listener who could critique without judging, and friends often left his company feeling understood—deconstructed and put back together again.

Once, just before Christmas, Jim watched John slip into a quiet conversation with a clerk at a craft store in Columbus, where they bought a German nativity set made of hand-carved wood. When they returned the next year to buy more pieces, Jim watched the clerk's face soften when John recognized her and picked up the conversation where they had left off twelve months earlier. Such a beautiful thing, Jim thought, to take the time to connect with strangers.

Still, the smallest thing could fluster John, a glance from a co-worker, a comment from a store clerk. On the last night of a trip to Paris in 1999, Jim stumbled as he tried to ask a waiter in French for a glass of red wine.

Je voudrais . . . de vin rouge?

"Oh hell," Jim said as the waiter stood rather impatiently at the head of the table. "How do you say *bottle?*"

"Bottle," the waiter said in perfect English, and promptly walked away.

John sank low in his seat, mortified. "Go outside," he told Jim. "You're embarrassing me."

Jim knew that John's heightened sensitivity was a vestige from childhood, when he was mocked by other children and berated by his father. Jim also knew that silliness could help defuse it, and so he walked out to the street and made funny faces through the brasserie's glass windows until the beginnings of a smile crept across John's face.

Early on, John decided that scenes of all distasteful varieties could be avoided with proper behavior, so he coached his friends to properly spoon soup (from the front of the bowl to the back) or match socks (never white) with slacks. He was a practiced host at lavish parties thrown in their cottage and later in an eight-thousand-square-foot house on a hillside overlooking Cincinnati's Mill Creek Valley, with a twenty-eight-mile stream running through the heart of the city and into the Ohio River just west of downtown. John and Jim filled each room of "the Big House" with paintings by local artists, and Jim found himself wondering whether John could have forged a successful career in interior design. He had been accepted into a program in Chicago when he was a senior in high school, but his father had refused to pay for it.

John even threw a party in the Big House in 2007 after his mother Marilyn died of lung cancer six months after she started coughing. Curtis had traveled with his mother on trips around the world, to Poland to study the activities of beavers, to Brazil to volunteer in an orphanage. In the final days of Marilyn's life, Curtis brought his laptop to the hospital and later to the hospice to sit with his mother and was the last person to see her before she died. But John could barely bring himself to go.

Mourning his sixty-four-year-old mother in public would have seemed almost obscene, so he and Curtis served a menu of Marilyn's favorite drinks at a party celebrating her life, including white wine "à la Marilyn," poured from a box and chilled with ice cubes, and put a blonde wig and some jewelry on a dress form that John found stashed in the basement.

"How are you doing?" Jim asked before the party, brushing his hand along John's cheekbone. John had cried once, just briefly, after the hospice nurse called with news of his mother's death. Jim had held him tightly, but he knew there would be no drama, no bucket lists, no urgent declarations of love or anger or grief. It was completely unfamiliar to Jim, who leaned toward sentiment, but after more than fourteen years with John, he had learned not to push his partner.

"I'm having a hard time, Boo," John said simply, then quit his job in project management and, over the objections of Curtis and Jim, bought a burned-out seven-unit apartment building in a desolate stretch north of downtown that he vowed to renovate for the good of Greater Cincinnati.

In 2001, when Cincinnati police shot and killed an unarmed nineteen-year-old African American man, prompting four nights of rioting, John decided to take part in a series of neighborhood discussions about race. He wasn't particularly political and would never think of himself as an activist, but he hated the idea that black people felt targeted and vulnerable, shunned by their own community. If he could help people sit down together and hear each other, he wanted to do it.

But redeveloping and renting out an apartment building would take years of hard work, and Jim feared that the complicated project would never get done. Curtis was so worried that he signed on as John's business partner to oversee the costs.

It came as no surprise to John's friends when he shrugged off a strange sensation in his left foot in the winter of 2011, a tingling that at first was more annoyance than disability. Meb Wolfe, who had married eleven years earlier, heard about it from her husband while she was away on a business trip. "John's having trouble with his leg," he said during a late-night call. "Something is definitely not right."

Katherine Jurs, a city planner who had met John in the early 1990s, got a text from Jim: John had a rough time getting home. I'm really worried.

Jurs thought about her longtime friends, who had opened their home year after year to fund-raisers she hosted for the Leukemia & Lymphoma Society.

Jurs responded: He needs to go and get checked out.

Jim: What could it be?

Jurs: A pinched nerve? It might not be as bad as you think.

At a Cincinnati bar one night, Jim watched quietly as John limped toward the bathroom. Their friend Jennifer Stowe, who had traveled with John and Jim to Paris, wondered whether John had suffered a mild stroke. "Look at him when he walks," Jim said, shaken.

"What's going on?"

"He's got to go to the doctor," Jim said. "He's tripped a few times, too."

It started with a strange sound, a soft and steady thump-thump-thump in the hallways of their condominium, which they had bought in 2008 after they'd decided it was time to sell the Big House and live a simpler life. John's left leg was slapping against the ground, as if bound by a heavy weight. *Thump-thump-thump.* The sound was deafening.

Then John had trouble swinging his legs out of their Volkswagen hatchback. He pointed to his shoes, brown leather lace-ups. "Too heavy," he told Jim with a shrug. They traded their hatchback for an SUV and chalked it up to benign aggravation, life in their forties.

But as the slapping grew louder, Jim trolled the Internet late into the night, reading about pinched nerves and lead poisoning and muscular dystrophy. One was worse than the other, but nothing was as frightening as a disease called ALS, a fatal neurological disorder that attacks the nerve cells in the brain and spinal cord.

In May, John went to the doctor. He started taking antibiotics for Lyme disease. He had blood infusions when traces of heavy metals were found in his body. He got acupuncture for a possible pinched nerve. Still, there was no firm diagnosis.

Jim knew that ruling out diseases like muscular dystrophy should have brought some comfort. But he also knew that an ALS diagnosis is more of a process of elimination, coming only after neurological exams and blood tests eliminate other conditions and disorders.

John had an MRI and spinal tap. The electrical activity in his skeletal muscles was tested. Still, by late May 2011, there was no diagnosis. Only questions.

Jim paced the kitchen, the bedroom, the living room, frowning at the white brick walls. He passed their painting of Tunisian kilns and the marble-topped coffee table that had once belonged to John's grandmother. He looked at the clock. He answered an e-mail from work. He waited. *Please. Let it be okay.*

Any moment now, John would come home from the neurologist with a final diagnosis, something that explained the heaviness in his left foot, which lately had been creeping higher, into his calf and thigh. One doctor had mentioned the possibility of ALS, which most often starts with weakening in a single limb, but other diseases could produce the same symptoms, and John wanted a second opinion. Hours earlier, he had dressed for the appointment, choosing a gray T-shirt with a picture of a drinking glass that said HALF FULL. Jim watched from the doorway.

"Let me come with you."

John paused and turned to Jim. "I need to go alone."

Jim fought the urge to grip John by the arms and beg him to change his mind. But John was about to find out whether he would live or die, and Jim couldn't think of a more personal moment. John would process the news on his own quiet terms, without emotion or drama.

Jim walked John to the door and said, "I love you," then sank down on a kitchen stool to wait. He looked around their condominium, on the second floor of a converted glass factory in a neighborhood of buildings from the mid-1800s. Their unit had been raw industrial space, but John and Jim renovated every room, adding white bookshelves and a backsplash by the kitchen sink of collectible pottery in orange, blue, and green. A painting of John and Jim from the 1990s hung on the wall in the living room.

Sitting in the kitchen, Jim thought about John's near-fatal asthma attack years earlier and promised himself that he wouldn't panic. The day before, Jim had sent an e-mail to his sister-in-law: We have our fingers crossed. He's been feeling better.

Jim brewed coffee and looked at the clock again.

When John walked in late that afternoon, he bent down to leave his leather shoes by the bench in the foyer, a sight so familiar and mundane that for a split second Jim felt himself relax. But John's cheeks were wet and his shoulders were slumped. Jim leapt from his stool in the kitchen, and under the foyer's stained-glass window, he gripped John by the arms. *Please. Let it be okay.*

There is never a good way to talk about dying, so John opted for simplicity. "Yeah," he said, the words a soft tremble. "It's ALS."

Jim wanted to scream. He wanted to vomit. How much time did John have? How quickly would his body fail him? John rarely cried, but his shoulders were heaving as he clung to Jim. "We'll do what we can, okay?" Jim said.

He took John's hand and they walked slowly down the stairs to the media room. The thump-thump-thump from John's heavy left foot seemed to reverberate off every wall in the house, and Jim silently cursed the disease that was crippling his partner. John leaned back into the couch, his face blotchy and streaked with tears. Outside, couples strolled the riverfront, relishing a first taste of summer. Soon, local storefronts would boast vibrant bundles of gladiolas and pansies. Jim felt disconnected and afraid, and he edged closer to John.

A veteran neurologist at the University of Cincinnati had made the diagnosis.

Back in medical school at the University of Illinois in the late 1970s, there was no question that John Quinlan would study the brain. He had been diagnosed with muscular dystrophy at fifteen, when he was playing on the high school football team, and decided he couldn't become a surgeon because he lacked the strength in his hands. By the time he was a neurology resident in the 1980s, he was navigating around the hospital on a power scooter. He took his last step in 2005.

Over the years, as his own body failed him, he diagnosed dozens of people with ALS, watching the muscles of their arms and legs waste away, leaving nothing but the contours of bone. A terrible disease, the doctor had said more than once, as patients who were once standing over his power wheelchair, looking healthy and strong, quickly lapsed into chairs of their own.

A slight man with a thick white beard and mustache, the doctor had decided that finding a cure for ALS must be like repairing a model of a miniature ship while it was inside a glass bottle. ALS not only affected the body, but the cells inside the body. It helped explain why the disease, after decades of research, had no known cure.

When Dr. Quinlan rolled into the exam room in his power wheelchair, notebook paper balanced in his lap, John was perched on the edge of the exam table. The doctor started asking John a series of questions. He wrote, "progressive weakness in left foot and left hand," "developed foot slap," "three falls," "can't hold a cup or clip fingernails."

Dr. Quinlan had seen the same symptoms in other ALS patients, but he still couldn't be certain. "Can you hold out your arm?" he asked John. "Don't let me knock it down."

John had trouble keeping his arm in the air.

The doctor scanned the results of John's medical tests, hoping to find another explanation for the muscle weakness. But he turned to John, and said, "I'm worried about this being ALS." Dr. Quinlan spoke slowly and carefully, knowing that patients often don't hear anything else after the shock of the initial diagnosis.

"I thought so," John said.

"What's your understanding of this disease?"

Most people diagnosed with ALS live for three to five years, but Dr. Quinlan had made it a habit of describing his own experience as a doctor: his longest-surviving ALS patient had lived for seventeen years, the shortest just nine months. There would be progressive muscle weakness, cramps, twitches, and slurred speech. Eventually, when John's breathing muscles weakened, he would need a ventilator. John had already decided he would never want a machine to help him eat or breathe.

"You've got a lot of living to do," Dr. Quinlan said. "Let's look at all the things that are important to you and the things that you want to do now."

That first weekend after the diagnosis, John and Jim let themselves sink into the comfort of a routine carved over years. It seemed there was safety inside their white brick walls, with John lounging in his corner of their favorite corduroy couch, long legs tucked beneath him. Jim couldn't bring himself to think about doctors and disease, not yet, not this soon, so he ironed shirts and scanned the *Cincinnati Enquirer* long into Sunday.

Then Monday came, swift and unwelcome. Jim knew that John considered the ALS diagnosis something of an embarrassment, as if he were responsible for the dying nerve cells in his brain. So that afternoon, Jim started making calls and sending e-mails to their family and friends, one after the next, feeling the contours of his life blur and fade every time he described the disease and their harsh new world.

He phoned Meb Wolfe, his older brothers, his sister Ann in Sandusky. He sent out an e-mail, struggling to find the words: I wish I had better news to share.

Paulette Roberts called from Portland, where she was vacationing along the Columbia River. Jim winced when he saw her cellphone number on his caller ID. "What's happening?" Her voice was urgent.

Jim hesitated before whispering the words he had come to detest. "Tootie, it's ALS."

There was silence, and then, "Oh Jim."

"I'm so sorry for telling you like this," Jim said, batting away fresh tears.

"No. No. You had to tell me."

"I'm sorry."

"There's no way you could have told me any better."

Paulette hung up, pulled the car to the side of the road alongside the river, and turned to her husband. "My God," she sobbed. "John's going to lose everything."

In the end, John decided to tell his brother, Curtis, but he did it the only way he knew how. Over an online bridge game, John e-mailed: I have some news—ALS.

Curtis, who was working in Saudi Arabia, e-mailed back: Do you want me to move home?

John e-mailed: It's okay.

Then, because there was a hand to play, John and Curtis finished their bridge game. Curtis went to work the next day and told his boss, "I just want you to know that I'm going to be funky for a while." Then he sat down at his desk and cried. Weeks later, he would transfer to Toronto to be closer to John.

John also called his father.

"ALS?" said Chester, who had been diagnosed with dementia. "That's Lou Gehrig's disease, isn't it?"

"Yeah, it is."

"Well," Chester told his son, "sometimes you've just got to play the hand that life deals you."

The first thing to go was their two-story condominium in the old glass factory. John and Jim had figured they would spend years here, in cozy corners filled with oriental rugs and antique furniture, perfect for a couple edging toward fifty. But John would need hallways wide enough for a wheelchair and windows big enough to let in the sun when he was no longer able to walk outside. In September 2011, they found a new condominium with an elevator in a bustling stretch of downtown not far from Cincinnati's famed Fountain Square, with its towering bronze statue, the Genius of Water, commissioned from Munich by a local hardware magnate in 1871.

They had bought three homes together over twenty years, but this time, Jim signed the deed alone. He tried not to cringe when John said firmly, "My name is not going to be on it."

That first winter brought a whirlwind of loss.

Left foot. Left leg. Left hand. John started wearing a leg brace and a metal plate in his shoe to control what doctors called "foot drop." He started walking with a cane, and when he complained, "I feel like everyone's looking at me, Boo," Jim encouraged him to buy one made of walnut with an arched, old-fashioned handle, something of a fashion accessory.

Years earlier, John had discovered a full-length men's mink coat while he was trolling a yard sale. He shrugged it on and said, "I like the way it feels." His friend, Adrienne Cowden, shook her head of brown curls and laughed. But John bought the coat and wore it when it was cold, and Cowden, a local pastry chef and historic preservation expert, realized that her sensitive friend could withstand stares from strangers as long as he could dictate when and how they happened. It was the stares he couldn't control that frightened him.

Left shoulder. Left arm. Left fingers. Jim started coaxing John's twitching arms into a dress shirt every morning before work. When John could no longer clasp the buttons, Jim took him to a tailor and had shirts made with Velcro instead of buttons.

Once, in those early months, on a trip to North Carolina's Outer Banks, it seemed the disease had taken a merciful turn. From a deck overlooking the Atlantic Ocean, John sipped beers as the chilly salt air whipped at his face. The sun was bliss, and for a perfect few days, the spasms and twitching disappeared.

But in the early, gray months of 2012, John's right foot started to drop and he traded the cane for a walker. There was no way to know what ALS would take next and there was no way to fight, no treatment to test. The human brain has a hundred billion nerve cells, which communicate with the muscles and glands through signals that can travel more than two hundred miles per hour, orchestrating even the tiniest of movements across the body. ALS kills the largest of the body's nerve cells, the upper and lower motor neurons, and science has never been able to say why.

Though descriptions of the disease date back to 1824, when a Scottish neurologist named Sir Charles Bell started studying the nerves in the spinal cord, research has been slow and interest from major drug companies sporadic. As neurological disorders go, ALS is relatively rare, with about 5,600 new cases a year in the United States, prompting the federal government to dub it an "orphan" disease.

When John was diagnosed, only one drug for it was on the market, approved for use back in 1995, but it prolonged survival by only a handful of months. Another drug was in clinical trial and appeared to slow the progression of the disease, a breakthrough treatment carefully watched by researchers, doctors, and ALS groups, but within two years, the trial would end with poor results, devastating the ALS community. Though there was growing interest among researchers and drug companies globally and advancements in quality-of-life treatments, it typically takes $2 billion and fifteen years of research to get a single drug to trial, and there was nothing nearly as promising on the horizon.

John had the fillings in his teeth replaced when he heard that metal might hasten the progression of the disease. He bought exotic mineral supplements and stopped eating gluten. But it seemed to Jim as if the disease was progressing far faster than he had imagined, and on the darkest days, when John struggled to hold a fork or brush his hair, Jim would sneak into an empty room, bury his face in his hands, and cry. Time had been tilted on its side, and Jim never knew when a new day might be the last with an arm that moved freely or a leg that could bend without pain. He lay awake long into the night.

John began to have trouble standing upright in public bathrooms and went to see Dr. Quinlan. The doctor pointed to a camera case that was dangling from the side of his wheelchair. Inside were wet wipes and a plastic container that could be used when urinating from a seated position. In April, John went back to Dr. Quinlan again, this time to learn about lifting systems that would help him into the car. Very soon, the doctor decided, John would no longer be able to walk.

For the most part, John rarely talked about the disease. When the *New York Times* cited the revitalization of Cincinnati's river front, John wrote on Facebook: "So happy and proud to live here."

John was working as a project manager, and Jim had taken a job with the same consulting company so he could be near John during the day. But typing had become a one-handed slog across the keyboard, and over dinner on a warm July evening, Jim asked carefully, "Do you need to think about going out on disability?"

"I really don't want to." ALS can severely affect the vocal cords, but John's speech, more than a year after the diagnosis, was still clear. "I feel like I need to keep working."

"Why?" Jim pressed. "You don't need the stress."

"I want to stay."

"You don't need to put yourself through this," Jim said. "Everyone will understand."

That week, Jim filled out the disability paperwork since John could no longer grip a pen. Jim could have taken a leave of absence from his own job, but as an unmarried man, he wouldn't have qualified for family medical leave. The idea seemed almost preposterous to Jim, who was regularly helping John use the toilet.

Years earlier, they had talked about marriage, particularly after Paulette Roberts called on John's forty-second birthday in 2007. "You can call me 'reverend' now. I got ordained," she said.

John laughed. "Do you mind if I ask . . . why?"

"Well, so I can marry you and Jim."

"One of these days, if we ever decide to get married, we'll have you marry us," John quipped. "But right now, we can't get married in Ohio."

Even if John and Jim traveled to a state that allowed gay couples to marry, the federal government only offered the rights and protections of marriage—more than a thousand benefits that covered everything from family leave to health insurance—to heterosexual couples. John and Jim thought, Why bother?

Many of their friends had consuming careers, but John and Jim considered their relationship the focal point of their world. And they already felt married. They were in love and committed for the long haul. They had bought houses, planned trips, saved for retirement. They had talked about adopting a child, but dropped the idea when they learned that Ohio didn't recognize two fathers. They hosted foreign exchange students instead, a boy from Sweden and another from Finland, with John issuing three rules: no drugs, no pregnancies, no riding in cars with friends who had been drinking. John and Jim took the boys on vacations and hosted dinner parties before their proms, creating a makeshift multinational family that remained intact long after the boys returned home a year later.

John and Jim weren't thinking about disease and death during those years, or the practical, critical benefits of marriage that are needed in times of crisis.

In the fall of 2012, John traded a walker for a manual wheelchair, which Jim hoisted into the back of their SUV on outings to dinners and concerts. John used the wheelchair in an ALS walk and helped raise more than $22,000 through a team that Jim dubbed "Half Full." Writing was slow and difficult, but when Cincinnati was cited as a leader in clean water technology in October, John posted on Facebook: "Another reason I love Cincinnati and why others should call Cincinnati home."

During a visit with Dr. Quinlan that month, John described his new life, navigated at just twenty-one inches off the ground. "Very upbeat despite severe disabilities," the doctor noted, adding, "Absolute need for a power wheelchair."

Jim started lifting John into bed, onto the couch, into the car. He had grab bars installed in the shower, and when John could no longer stand on his own, Jim went out and bought a shower seat. Eventually, Jim stripped down and got into the shower himself to hold John steady. The last limb to go was John's right arm, and with it came a power wheelchair that Curtis quickly dubbed the "urban assault vehicle" in a halfhearted attempt to make John laugh. John hated the chair's bulk and the embarrassing hum it would make, and he told Jim just before Christmas in 2012, "I don't want to go out in that thing. I feel like a spectacle."

Then John had to trade the wheelchair for a bed, and it seemed the only thing left to do was wait.

One of the cruelest ironies of ALS is that its victims lose their muscles and motor skills, but not their ability to feel pain. In John's case, it often shot down his legs, hot and unrelenting, and Jim spent hours hunched by John's bedside, shifting his position to relieve pressure. Jim brought in a hospital bed, set it up in the center of the bedroom, and moved himself to the guest room across the hall.

Paulette came by to visit and sucked in her breath when she saw her nephew, huddled under an electric blanket. "You know I'm going to die, Aunt Toot," John said when Jim was out of earshot. "It's going to be months, not years. I just want to tell you it's coming."

Paulette thought of the skinny, sensitive boy who would crawl into her lap for songs or stories as she looked at the man in the hospital bed. She kissed John on the forehead, careful not to press too hard. "It's okay, sweetheart," she said. "I know."

To Jim, the call to hospice seemed almost surreal, even after months of living with the menace of a deadly disease. But nearly every limb on John's body had lost function and his speech had started to go. Jim felt as if he was shifting between two extremes, numbness and absolute misery, when he e-mailed friends in March 2013 to say that John could no longer answer calls on his cell phone.

He wrote to his sister, Ann: This f'ing disease.

Martha Epling, a family friend, had started working at a local hospice three years before she lost her seventy-four-year-old husband to pulmonary disease. With soft white hair and red glasses, she reminded Jim of what it might have been like to talk to his own mother.

"You need help," Epling said, studying Jim over a coffee mug in a downtown café that spring.

Jim had been staying up at night, worried that John would need the bedpan changed or his position shifted in bed. "I have to do it myself, Martha."

"No, you don't." There were nurses, social workers, physical therapists. "You don't have to be his total entertainment. You don't have to do all of the physical stuff."

She went to see John a few days later while Jim was at the grocery store. "Will I be comfortable with strange women giving me a bath?" John asked her, trying to smile. Then he said, "I need to do this. I need to do this for Jim."

With nurses coming twice a week, Jim decided to continue their tradition of turning Friday nights into something of a celebration. He invited friends to sit by John's bedside with wine and Manhattan cocktails. One night, their old friend Meb Wolfe stopped by for a visit.

"I hope you don't mind us just hanging around," she said, huddled over John, who had a bowl of guacamole balanced in his lap. The tortilla chips were large, and John could still manage to draw one into his mouth.

"No. This is great," John said, glancing around the room. "This is my world now."

Wolfe couldn't think of a single thing to say. She looked at her friend, trying to drink a splash of bourbon from a cup with a straw, and thought, I can't believe we're even having this conversation.

By May, John's arms stopped moving altogether. His neck started to ache, even after a drink of water. Jim wanted to stroke his hand, rub his head, run a finger along his cheekbone, brief moments of intimacy that made Jim feel as if he was still connected to his partner. But the disease took that away, too. Jim was profoundly sad and too tired to think clearly. When their traveling companion, Jennifer Stowe, came by, John gestured toward Jim and whispered, "Please. Get Jim out of the house."

Jim didn't want to leave, but he agreed to step out to a neighborhood pub with Stowe. She studied Jim, who had grown a beard and gained twenty pounds. Jim had stopped cooking because John couldn't eat, and he often stole downstairs to a donut shop on the ground floor of their building. Over tacos, Jim yawned. "I'm sorry I'm not much fun. I just spent most of the day crying."

"I don't know how you can't," Stowe replied, thinking about the twenty-eight-year-old John Arthur who had nicknamed her "Fire" when he discovered that her Ford Mustang had gone up in flames in college after a botched oil change. All through the years, on trips to London, Paris, and the North Carolina Outer Banks, the nickname stuck.

Stowe leaned over the table and kissed Jim's cheek, but the gesture seemed entirely inadequate.

"I feel awful," Jim said. "Am I doing enough?"

"You're doing plenty. You're there. You're there every day and you see it."

"It's just that you can't do anything about it," Jim said slowly. "You just have to sit and watch."

11

"I DO"

ON JUNE 26, 2013, two years and one day after the ALS diagnosis, John called softly from the bedroom. "Jim, the Supreme Court ruling is about to come out."

Jim had been up since dawn, answering e-mails from work at the dining room table, and he was already bone tired. He slept in the guest room but had a wireless doorbell set up next to John's hospital bed so he could ring for help in the night when he needed his position shifted. Soon, Jim knew, John would no longer be able to push the bell, and they would swap it for a pressure-sensitive pad that buzzed when John inched his head slightly to the right.

Jim hurried to the bedroom, where John was propped up under a blanket. On television, MSNBC's Chris Jansing said, "Take a look now, huge crowds outside the Supreme Court today. Nine justices will rule on two cases that could redefine equality and gay rights in this country."

Jim reached for John's hand, eyes fixed to the screen.

For weeks, they had been waiting to find out whether the Supreme Court would strike down the seventeen-year-old Defense of Marriage Act, known as DOMA. President Bill Clinton's press secretary at the time had called the law "gay baiting, plain and simple."

John and Jim had been a couple for just four years when the law passed, denying same-sex spouses access to federal marriage benefits. DOMA also allowed states to refuse to recognize same-sex marriages even if they were legally performed in other states, a provision that one day would alter the course of Jim's life long after John was gone.

The eighty-four-year-old plaintiff at the center of the Supreme Court case was the daughter of Russian Jewish immigrants who had lost the family's ice cream shop and home during the Great Depression. In 1965, young Edie Windsor, working at IBM, started dating Thea Spyer, whose family had fled the Nazi invasion of the Netherlands. After forty years together in New York City, they decided to marry in Toronto, Canada, where same-sex marriage was legal. Spyer died from a heart condition two years later. Though federal law allowed a spouse to leave assets to a surviving spouse without incurring estate tax, the IRS sent Windsor a $363,000 bill, as if the women had never been married.

The case seemed outrageous to Jim, who wondered over long discussions with John how a grieving woman could be treated with such dispassion. If the Supreme Court required the federal government to respect the lawful marriages of same-sex couples equally under the Constitution, it would be the biggest victory in the ninety-year history of America's gay rights movement, which had started in Chicago with a fledgling organization for gay men called the Society for Human Rights.

In the bedroom, Jim glanced at John, who was watching the broadcast intently, his dark glasses perched on the end of his nose. It was a bright, 90-degree day in Washington, D.C., and before the U.S. Supreme Court's towering Corinthian columns, flag poles and fountains, kissing couples waved American flags and signs that read ALL LOVE IS EQUAL. NBC's Pete Williams announced over the cheers, "The Supreme Court has just struck down the federal Defense of Marriage Act."

Jim sucked in his breath and bent down to embrace John, happier than he had been in months. The broadcast continued, " . . . The federal government can't make distinctions between same-sex and opposite-sex couples in terms of what marriages the federal government would recognize."

The Supreme Court had also declined to rule on the merits of Proposition 8, a voter initiative that banned same-sex marriage in California, reinstating a lower court ruling that had allowed the state to issue licenses to gay couples.

Jim kissed John lightly, loving the feel of his face, and then, because it seemed as if the country was changing, Jim looked at his dying partner and said for the first time in twenty years, "Let's get married."

John said yes.

The *Windsor* ruling only addressed whether the federal government had to recognize the existing marriages of gay couples, leaving open the question of whether states like Ohio, with longstanding marriage bans, had to respect them as well. But the U.S. government for the first time would broaden its definition of marriage to include the single most important relationship in Jim's life, and for now, that was enough.

Jim raced to his computer in the dining room. At 10:56 A.M., less than an hour after the Supreme Court announcement, he e-mailed Paulette Roberts:

> We may need your services. As soon as the Supreme Court ruled DOMA unconstitutional, I hugged John and asked him to marry me. That means getting to the closest state

> that allows gay marriage but doesn't require residency. That's New York. Can you marry people in New York? I can't imagine having anyone else marry us.

Suddenly impatient, Jim decided to call. "What are you doing next week?" he asked Paulette, laughing and crying. "How about a trip to New York?"

Paulette hadn't yet heard about the decision, but she could hear the giddiness in Jim's voice. "What happened?"

"I asked John to marry me and he said yes."

Paulette didn't hesitate. "Wherever you want to go."

She called a court clerk in New York City who shouted into the phone, "Yes!" when Paulette told her about that morning's Supreme Court ruling. Two years earlier, New York had become the seventh state in the country to allow gay couples to marry. When the legislation passed, Niagara Falls was lit in rainbow colors.

The problem, Jim quickly learned, was that New York required both partners to show up at the courthouse and apply for a marriage license, then return for the ceremony. John's last trip out of the house four months earlier had caused great discomfort, and that was only a three-mile trek across town in their minivan. Jim had to find another way.

He scrambled to research the rules in other states and immediately settled on Maryland, which required only one person to appear at the courthouse for a marriage license. In the 1970s, Maryland was the first state in the country to pass a law that expressly defined marriage as a union between a man and a woman. Forty years later, it was among the first to extend marriage rights to same-sex couples by popular vote, a watershed moment in the gay rights movement.

Jim would fly alone to Baltimore and then drive to nearby

Annapolis, sprawled along the Chesapeake Bay thirty miles east of Washington, D.C. But he still had to get John to Maryland for the ceremony. He looked into ambulances, which were too slow and bumpy, and was more than once struck by the absurdity of having a bedridden man travel 520 miles when the Hamilton County Courthouse was only six blocks away. The best option was a plane fitted with medical equipment, which would get to Maryland quickly and had room for a stretcher. The flight would cost nearly $14,000.

Their story moved Julie Zimmerman, a former neighbor and writer who was on the editorial board at the *Cincinnati Enquirer.* She set up an interview with John and Jim and brought along video producer Glenn Hartong. John couldn't speak a sentence without pausing to breathe, and his words were slow and halting.

"Had the Supreme Court made this decision one year earlier," he told the journalists, "this would have been as simple as us taking a trip because I could still walk. It's the progression for me of ALS. It's just compounded everything."

"I don't live in a world of regret." John said carefully. "But I sure wish we were a year in the past."

There was a long pause, long enough for Hartong to pan, reframe and refocus his camera.

Finally, Jim smiled gently and said, "Me too."

Watching the exchange, Hartong thought it was the most powerful understatement he had ever witnessed as a journalist.

Relatives and friends donated for the medical flight, and Paulette offered to stay with John while Jim flew to Maryland to apply for the license. Even twelve hours away from home required significant planning, but there was simply no time, so Jim rushed Paulette around the house, to John's stash of pain medication, to the controls on his airbed, to the blankets, towels, cups, computer.

Paulette would remember how John politely refused to eat or

drink that day, clearly trying to stave off the need to use the bedpan. With no ability to move his arms or hands, he would need help positioning himself. Finally, John had said, "Would it bother you to touch me?"

Damn this disease, Paulette thought, wondering how long he had waited before asking. "John," she said, pulling back the blanket. "The only reason I'm not touching you is because I know it hurts."

Jim arrived home late that night. Two days later, after the Maryland marriage license came in the mail, he posted on Facebook: "One step closer to being a married man!"

Two pilots and a nurse were waiting on the tarmac when the ambulance rolled into Cincinnati Municipal Lunken Airport just after nine A.M. and came to a stop next to the medical plane that would fly John and Jim to Maryland.

John, wearing a green shirt with Velcro that he'd bought when he had still been able to walk, was buckled into a narrow gurney. The ambulance ride had been uncomfortable, and Jim clutched John's hand as a couple of paramedics flung open the back doors. The sun glared off the waiting Learjet. "I can't believe it's happening," Jim whispered to John.

John's eyes were closed, but he opened them when Paulette jumped into the back of the ambulance and kissed him on both cheeks. Jim wiped John's mouth with his fingers, then stepped out onto the tarmac with his backpack. Tucked inside was the marriage certificate with the state of Maryland's official gold seal: *James Robert Obergefell and John Montgomery Arthur.*

Two paramedics heaved John's gurney out of the ambulance and maneuvered it through the plane's narrow doorway with help from one of the pilots. Jim hugged Martha Epling, who was waiting on the tarmac with a hospice nurse, and patted his backpack.

"One last check, make sure I have the all-important marriage license," Jim said, digging through the bag. Then he boarded the plane with Paulette.

The takeoff was smooth, and Jim sat back in his leather seat next to John's gurney. He gripped John's hand again, worried that the stress of travel would escalate the disease's relentless crawl across John's body. Jim could see that the bars on the gurney were digging into John's back and legs, and Jim hated how John suffered. But John slept as the plane soared toward Maryland.

This day, more than any other, needed to be perfect. Jim thought about the ring that John had given him on that rainy night in a Columbus hotel in 1993, with five perfectly set diamonds. Jim had worn it off and on for years, but now he would wear a wedding ring cast in sterling silver. Staring idly at the wisps of clouds just beyond the plane's windows, it seemed to Jim as if each day from the past twenty years had led them to this moment, and he felt a rush of tenderness as he looked at John.

More than once in recent months, John had said, "I feel guilty about ruining your life." John saw marriage as the only way of making certain that Jim was cared for financially and legally. Jim just saw love, profound and enduring, and a marriage ceremony would allow him to tell John that.

Ninety minutes later, the plane touched down on a runway outside of Baltimore. They would marry here, inside the plane's tiny cabin, in front of Paulette and video producer Glenn Hartong, who had flown to Baltimore earlier that morning to meet them. There was no room for standing, but John was on a stretcher anyway, so Jim leaned forward in his seat and absently brushed his thumb back and forth across John's limp hand. John fixed his eyes on Jim.

Paulette sat behind them and said, "Today is a momentous day not only in the lives of two of the most loving special men I have ever known, but also in the lives of all who know, love, and respect them. . . . Twenty-six months ago, John was diagnosed with ALS. Since then, the amazing relationship between John and Jim has become even closer, even more devoted, even more loving.

"And it was pretty damn great before John became ill."

John managed a half smile.

Jim had written his vows a few days earlier, surprised at how easily the words flowed. He said softly, "We met for the first time. My life didn't change. Your life didn't change. We met a second time. Still nothing changed. Then we met a third time, and everything changed."

He cleared his throat. "As you recently said, it was love at third sight. And for the past twenty years, six months, and eleven days, it's been love at every sight. You've taught me generosity. You've taught me balance. You've given me joy. You've loved me when it was easy and when it was difficult. You've made me a better person. Thank you—for seeing in me someone you could spend your life with. Thank you—for allowing me to love you when you thought you were lost and beyond love."

"I give you my heart, my soul, and everything I am," Jim said. "I am honored to call you my husband."

Jim reached for John's left hand and pulled it close so he could wiggle the ring onto John's finger. Jim vowed, "With this ring, I thee wed." Then he lifted John's left hand, moved it toward his own and slipped his own ring onto his finger.

"With. This. Ring. I. Thee. Wed," John said, careful not to trip on the words.

"John Montgomery Arthur, do you, continuing from this day, take James Robert Obergefell to be the love of your life, your eternal partner, your husband?"

"I do," John said.

" . . . In witness of those present, in the loving thoughts of the many people who wish they could have been with us today, in the spirit of loved ones who have come before, by license granted to me by law and with the respect of the law by our great land, I now pronounce you husband and husband, forever intertwined partners. May love and good will be with you forever."

Jim kissed John. Paulette leaned forward and hugged them both. "You're beautiful," she said, weeping.

John responded: "Thank you. And thank you for including the word *damn.* . . . "

Paulette and Jim were still smiling when the plane took off for home just after eleven A.M. In Cincinnati, a small crowd waited on the tarmac with signs that read FINALLY: MR. AND MR. Jim could hear the cheers as soon as the hatch opened, and he grinned and ducked through a shower of white rice. "I'm overjoyed, in love, and so thankful," he said.

When the paramedics lifted John's gurney out of the plane, Jim leaned over and kissed him on the forehead. At forty-six, Jim was a married man with rice in his hair and a wedding ring on his finger, and he momentarily forgot about the disease that was killing his husband. He looked at John, who said, "I'm very proud to be an American and . . . to openly share my love for the record, and I feel like the luckiest guy in the world."

For five days, married life was a blissful burst of words and phrases that John and Jim had never been able to use before.

"Good night, husband," Jim said as he settled John into bed.

"Hey, husband," John said in the morning. "I'm thirsty."

Jim changed his Facebook status to: "Married to John Arthur."

He opened a bottle of champagne when friends stopped by and slipped some into John's sippy cup. "Some more, husband?" Jim said. The phrase shimmied off his tongue. There would never be a honeymoon, but after two decades of words that had always sounded imprecise—partner, better half, significant other—this was enough.

John would die a married man, and Jim would grieve a husband.

At a dinner party across town a few days after the wedding, Al Gerhardstein ran into an old friend, Barbara Cook, a lawyer who represented veterans with disabilities. Cook and her husband had lived for years near John and Jim when they owned the Big House and had cruised parts of Canada with them a year earlier, playing rowdy trivia contests in the ship's bar as John piped in from his wheelchair.

Over dinner, Cook told Al and the group about the wedding, which had been posted on the *Cincinnati Enquirer*'s website and was gaining interest on the Internet. "You have to watch the video. It's just so moving."

"Wait a minute," Al said. "They got married in Maryland and now they're back in Ohio? Do they know the law?"

"I'm in tears over this and you're talking about the law?" Cook said, poking fun at her friend.

"Right. Right," Al answered, and then he slipped away. Cook found out later that he left the party, went straight home, and started researching case law.

To take on Ohio's marriage recognition ban, Al needed three things: a precedent, a plaintiff, and a story.

He had found a precedent when the Supreme Court in *Windsor* declared a key section of the Defense of Marriage Act unconstitutional, forcing the federal government for the first time to recognize same-sex marriage and deliver all the benefits provided to heterosexual couples. Al had pored over the decision once, twice, wondering whether the ruling gave him an opening in Ohio. Why should states follow different rules than the federal government?

He met with Lisa Meeks, who had been one of a handful of openly gay lawyers in Cincinnati in the 1990s and had built a practice helping lesbian and gay couples with estate planning, domestic relations matters, and employment and housing issues. Later, Meeks started drawing up shared custody agreements for same-sex couples since Ohio only recognized the biological parent and deemed the other a legal stranger. In the case of an Ohio adoption, only one parent had legal status.

"There are more ramifications to not being allowed to marry than just not being allowed to marry," Meeks told Al. "You can't really create or protect your family. It's more than just symbolic."

Al nodded. "Let's think about this."

Five days after the *Windsor* ruling, Al e-mailed Meeks: Do you know couples who are married in another state but living here in Ohio?

Meeks said yes.

Six days after the ruling, Al e-mailed: If a same-sex couple from a state where same-sex marriage is legal adopts a child in that state and moves to Ohio, does Ohio recognize both of them as parents?

Meeks said no.

Sixteen days after the ruling, Al e-mailed again. Do you know of differences in legal age or anything else that Ohio ignores when heterosexual couples arrive in Ohio, already married?

Al and his law clerks quickly researched state policy, finding that Ohio had long banned first cousins and minors from marrying but would recognize such marriages when they had been performed legally in other states. That wasn't the case for same-sex couples.

Seventeen days after the *Windsor* ruling, Al learned about Jim Obergefell and John Arthur, who had traveled all the way to Maryland to marry even though John was bedridden.

Al e-mailed his law partner, Jennifer Branch, to explore whether they had a basis for an injunction—an injury so great that the court would need to act immediately to avoid "irreparable harm." He wrote: When a same-sex partner dies in Ohio, I assume the box on the death certificate for marital status is checked, "never married," and the name of the surviving spouse is blank, right? Isn't that irreparable harm that your legacy, your family, will not be noted ever in Ohio?

Across town, Barbara Cook reached out to Jim. She e-mailed: A coincidental bit of timing: a civil rights lawyer friend of mine . . . thinks that a lawsuit is the only way the ban on gay marriage will go away here.

Cook wasn't sure if Jim would be interested in a lawsuit since he was newly married and busy caring for a dying husband. But nine minutes later, Jim responded: I want to do what I can.

Cook replied: So, Al, meet Jim; and Jim, meet Al.

Al had his precedent, his plaintiffs, and his story, and just before he filed suit in federal court, three days after meeting Jim and John, twenty-three days after the Supreme Court's *Windsor* decision, and fifteen years after the devastating defeat on Issue 3, he went home and called his daughter Jessica, who was about to start law school at the University of Michigan.

Jessica liked to tell the young men she dated, "My dad is the best person on earth." But she also knew her father could be impatient, hard-charging, and anxious about the prospect of disappointing his clients.

"I don't know if I should do this," Al told his daughter.

It was clear to Al that Cincinnati was changing. The city had hate-crime and anti-discrimination laws and, two years earlier, the first openly gay person was elected to city council. Twenty-three candidates had competed for nine seats, but at age thirty-one, Chris Seelbach had secured enough votes to win the last spot. Though Chris had spent years struggling to help his parents understand that he was gay, his father on Election Day had held a sign in front of a busy northwest precinct that read PLEASE SUPPORT MY SON.

Late that afternoon, an angry man had approached him and quipped, “The last thing I need is a queer city council member.” Steve Seelbach, with a linebacker’s build, chased the man back to his car.

But beyond the borders of Hamilton County, Al knew that antigay sentiment still ran strong, with polls showing that about 40 percent of Ohio supported the state’s nine-year-old ban on same-sex marriage.

“I feel like I’m cursed on these issues,” Al told Jessica.

“No,” Jessica urged. “Do this.”

PART THREE

LOSS

"Law and order exist for the purpose of establishing justice and . . . when they fail to do this they become the dangerously structured dams that block the flow of social progress."

—Martin Luther King Jr.

12

DUTY TO DEFEND

IT WASN'T exactly an enviable job, challenging a dying man, but Bridget Coontz was an expert at constitutional law and had been dispatched by the Ohio attorney general to defend the state against the lawsuit filed by John Arthur and Jim Obergefell.

She was only three months into her new position at the attorney general's office, charged with supervising fourteen attorneys who represented the governor and state agencies when lawsuits were filed against them. On Friday, July 19, 2013, Coontz's supervisor had hurried over to her office and said, "Al Gerhardstein filed a request for an emergency injunction. Same-sex marriage in the context of death certificates."

With a degree from Rutgers School of Law, Coontz had sued the slumlords of Columbus, Ohio, for letting their buildings rot without basic maintenance. She had trained state highway troopers on the proper way to conduct searches and seizures and attorneys on the laws of free speech. But she knew that no other assignment would draw more attention than a case on same-sex marriage.

She looked at her anxious boss and said, "I'll take it."

It was already four P.M., which meant she had just over two days to prepare for the emergency hearing that Judge Timothy Black had scheduled for the following Monday. Her weekend was packed with family activities with her husband, a police officer, and her seven-month-old twins and four-year-old daughter. But the bigger problem was more philosophical than practical: Coontz supported same-sex marriage.

She had friends who were gay and saw no compelling reason why they shouldn't have the right to marry. Now, as the counsel of record in a lawsuit against the State of Ohio, she would have to walk into federal court and defend the way the state would handle the last record of a gay man's life. More than once over the weekend, she thought about introducing herself on Monday morning by saying, "Bridget Coontz on behalf of the wrong side of this courtroom."

But she knew the attorney general believed in the "duty to defend" laws that had been passed by Ohio voters, and she was grateful when she learned that the state would argue only for the democratic process—not against gays. Though the attorney general had said publicly that he personally opposed same-sex marriage, no psychologists would be called in to discuss the attributes of traditional marriage or whether gay couples made suitable parents.

"We are not going to crazy town," she and her colleagues said more than once. "Our arguments are strictly legal."

Early Monday morning, she kissed her husband and children, ran a brush through her orange hair, and put on a suit. *I can defend my client,* she thought, *but still control the message.* She would argue that there was no need for an emergency order because the court could always require that John Arthur's death certificate be changed later, after the judge held a final hearing on the matter.

Sitting in court, she listened to Al Gerhardstein plead for an emergency order and watched Jim Obergefell take the witness stand to read a statement about his marriage to John. Glancing at Al and Jim at the plaintiff's table across the room, the assistant attorney general stood up and said, "This is a sympathetic case and it's a hard one."

" . . . Now, plaintiffs are before this court asking it to effectively, but only temporarily, do something that no other federal court has done in an overnight, temporary restraining order, and that is strike down a state statute and constitutional provision, which defines marriage as a union between a man and a woman."

Judge Black was silent. "Perhaps the most important factor in this case is the lack of irreparable harm," she continued. " . . . By statute, a death certificate can be changed; therefore, the harm that plaintiffs allege, an omission on the death certificate, an omission related to marital information, is, by statute, reparable."

Judge Black cut in a moment later. "What if the plaintiff dies in the meantime?"

"I'm sorry?"

"What if the plaintiff dies in the meantime and then we go back and fix the death certificate, if that's what the law requires. Has not that dead plaintiff suffered irreparable harm?"

"The harm being that the plaintiff died with the death certificate that did not reflect their marriage?" Coontz asked.

"Correct," the judge replied.

"Your Honor, in that situation, it's the knowledge that we're talking about. The plaintiff is going to—if he passes away—" she started.

Again, the judge interjected. "He's going to pass away."

"When he passes away," Coontz continued, "he's not going to know what his death certificate says. Unfortunately, nobody does. Nobody dies with the knowledge of what their death certificate says. And if, ultimately, a death certificate is incorrect, Ohio law carves out a way to fix it."

She pressed on. "... The balance of harm in this situation weighs in favor of, once again, slowing this case down. While we understand that there's the possibility that the plaintiff could die in the meantime, what we're talking about is the recognition of one marriage. . . ."

My marriage, Jim thought.

A few minutes later, Coontz sat down. She had made her case, defended her client. It was done.

Judge Black called on the last lawyer at the hearing, the deputy solicitor for the City of Cincinnati, who had come to court because the local health department, charged with finalizing death certificates, had been named a defendant in the case. The city, in every sense, was caught in the middle, compelled to comply with Ohio law, which banned the recognition of same-sex marriage, even though Al Gehardstein was alleging the law was unconstitutional.

Before the hearing, top city officials had come to a decision about how they would to respond to the lawsuit. Now, city attorney Aaron Herzig stood up, looked at Judge Black, and said, "I think the actions of the City of Cincinnati, over the last decade or so, indicate that the city is very sympathetic to the idea that Ohio should be a place where same-sex marriage is permitted."

"The City of Cincinnati takes the position in this litigation that it will not defend the propriety of the same-sex marriage ban in Ohio?" the judge asked.

"That's correct, Your Honor," Herzig replied.

Jim's hands started shaking again. John had once taken an urban planning class at the University of Cincinnati, studying the decline of American cities, and had promptly vowed to never move outside Cincinnati's city limits. He had collected artwork from Cincinnati artists, supported Cincinnati charities, bought houses in Cincinnati neighborhoods. Listening to Aaron Herzig break ranks with the state, Jim silently thanked John. Damn, Jim thought. Cincinnati is on our side.

Judge Black was looking at Al. ". . . If the court were to put on a temporary order, is not the state harmed when a single federal judge with the stroke of a pen puts on hold a democratically elected amendment to the Constitution?"

Al had prepared for this question. "Judge, this is how our system works. And the worst way to protect minority rights is to put them up for a vote."

"The mere fact that voters establish law doesn't make the law lawful?" Judge Black asked.

"No."

"I will act today," Judge Black said. "Give me the time to do what the evidence and the law requires."

From her seat in the middle of the courtroom, John's aunt Paulette thought about Al's parting words. *The worst way to protect minority rights is to put them up for a vote.* Surely, Judge Black would see it that way, too. "I think I know which way he wants to go," she told her husband, Mike, a retired Cincinnati lawyer, "but I don't know how he ultimately has to judge it."

Neither did Al, who had made it a habit over the years not to offer predictions about the outcome of a case. He had met Tim Black in the 1980s when the judge, then a practicing attorney, was on the board of Planned Parenthood during the protests and bombings. But that was nearly three decades ago. Since then, Tim Black had run for the municipal court, first as a Republican and subsequently as a Democrat, and on the federal bench, he was considered tough, professional, and precise. In the hallway outside the courtroom, Al told Jim, "We'll just have to wait and see."

Jim, Paulette, and Mike went back to the condominium to be

with John, who was waiting for news about the hearing. "We don't know how it's going to go, but it's going to happen today," Paulette told her nephew.

John smiled from his hospital bed. "I hope so."

Television crews were waiting on word of a decision at the courthouse. "If you want to go back," Paulette told Jim, "we'll sit with John."

Jim took John's hand. "No, I want to be right here."

They waited.

Tim Black was the son of a Boston heart surgeon, but he had always leaned toward the law. His great-grandfather had spent twenty-six years as a trial court judge in Cincinnati and his uncle served on the Ohio Court of Appeals. "If you treat even the most desperate man with dignity and respect," his uncle once told him, "he will accept your judgment."

The sentiment had stuck with Black, who had made it a practice to ask law students, "What's the most important trait in a judge?" He got the same answers nearly every time: fairness, wisdom, integrity. But the judge would cite Solomon, the ancient king of Israel, who spoke of a kind and understanding heart.

"When we study the biographies of our heroes," the judge wrote, "we find that most of their lives was spent in quiet preparation, doing tiny decent things, until one historic moment catapulted them to center stage."

After the hearing, three years into a lifetime appointment to the federal bench, he walked back down the hall to his office, thinking about the case of Jim Obergefell and John Arthur. The judge's office was cool and quiet, filled with photos of the Caribbean islands he had visited on sailing trips with his family. He had been five when he started racing sailboats and in his teens when he started winning championships. In his early twenties, he started out on a yearlong journey to New Zealand with his wife and another couple. The wind and waves off the coast of the Bahamas had been terrifying, so strong that they ripped the boat's main sail and flooded the pump, and the two couples spent the night desperately bailing water from the boat with buckets. They decided not to cross the Pacific and instead visited nearly every island in the Caribbean.

He planned to sail again that upcoming weekend at Cowan Lake State Park near Cincinnati. But now the judge had a decision to make, one that would surely draw attention from the media. He had read Al Gerhardstein's complaint and the state's response and couldn't find a single good reason to deny John Arthur a death certificate that described him as a married man. Fundamental constitutional rights, he believed, could not be subject to the vote of the electorate.

The judge had already started drafting a preliminary ruling, starting with the words "This is not a complicated case." Alone in his office, he made a few final revisions and turned in his decision at 5:01 P.M.

Minutes later, across town in Jim and John's condominium, the phone rang. Jim was still sitting beside John, who had dozed off and on throughout the afternoon. Jim glanced briefly at Paulette and Mike, waiting expectantly on the other side of the bed.

"Hello?"

"Jim. It's Al."

Jim had known Al for less than a week but had already picked up on his lawyer's habit of delivering news with controlled, measured words. The only place Jim saw emotion was in Al's eyes, when Al had looked at John in the hospital bed or at Jim on the witness stand, describing what marriage had meant after disease had taken everything else.

"Jim," Al said. "We won."

Jim hung up the phone, breathing hard. *Holy hell. We did it.* He thought briefly about Edie Windsor and all the other same-sex marriage plaintiffs who had come before him, in states like Hawaii, California, Massachusetts, and Vermont, where lawyers and couples spent years fighting for the right to marry. Then he bent down to rouse John.

"I'm not sure what it was," Jim said, "but it went our way."

Paulette threw her hands in the air and grinned.

13

GOOD NIGHT, HUSBAND

DAVID MICHENER hadn't heard of the federal court ruling the night his husband was rushed into emergency surgery to stop a bacterial infection that had mauled the right half of his heart. It could have been a paper cut, the doctors said, that caused the infection that caused the fever, sending fifty-four-year-old Bill Ives into intensive care with a failing aortic valve in late August 2013.

Still, Michener fully expected his husband to survive, in part because he had to. The adoption papers had come earlier that day, and three-year-old Michael, who had faced a childhood in foster care, was now officially their son. The toddler and the couple's two older children were waiting at home in a ranch house in the historic Cincinnati neighborhood of Wyoming, known for its top schools and panoramic views of the city. Michener, broad-shouldered and a good head taller than his husband, was a business analyst who liked to pen poetry. He could sit out back for hours, watching deer and cardinals dart between the trees.

Michener and Ives were newlyweds, married just five weeks

earlier at their beach house in Delaware. But they had spent nineteen years together in the suburbs of Philadelphia, living a rather ordinary life that started with a commitment ceremony, then a house, then plans for a family. They had raced to New Orleans to pick up their adopted daughter, Anna, a two-pound preemie whom Ives was afraid to hold until Michener coaxed him into it, then went back two years later to get Jack, born during a hailstorm that shut down the highways of Baton Rouge. The couple commandeered a police escort to get to the hospital.

In 2008, they moved to Cincinnati so Ives could take a corporate job at Macy's, but nearly turned back when they crossed into Ohio, where billboards delivered word of the Bible. "We made a mistake," Ives said in a panic. "We're in God's country."

"Relax," Michener said. "We've always lived an out life."

In Pennsylvania, they had taken the kinds of precautions common among unmarried gay partners with adopted children, drawing up parenting agreements and wills. Married couples had automatic rights, but unmarried gay couples had been refused access to their dying partners, inheritances, and nonbiological children. In Ohio, the couple would hire attorneys Lisa Meeks and Scott Knox to draw up more paperwork, including a joint custody agreement that had become a fallback for gay couples in a state that refused to allow same-sex parents to jointly adopt children.

Michener decided to stop working to stay home with Anna and Jack, juggling gymnastics, tennis, and martial arts competitions. Then, at a family meeting in early 2012, they decided to adopt Michael, an African American toddler who clung to his new fathers when the social worker brought him over for his first stay with the family.

In July 2013, when Al and Jim were talking about a lawsuit, Ives and Michener traveled to Delaware, the eleventh state to allow gay couples to marry, and exchanged wedding vows in front of their children. But Ives was nursing a 104-degree fever, and though he believed it was a lingering flu, they drove home the next day for medical tests. He was admitted to the hospital for three weeks of intensive antibiotic treatments.

"If anything were to happen, this is what you have to do," Ives said from his hospital bed, describing their estate, their wills, and the calls that would need to be made to lawyers.

"Just shut up," Michener replied. "You're not going anywhere."

But Ives didn't get better, and on August 24, just after Michener dropped his children at the bus stop, Ives was rushed into surgery. Michener drove straight to the hospital, thinking about how his husband had laughed from the hospital bathroom just two days earlier. "Oh my God," Ives had said, looking at his gray hair in the mirror. "Look how f'ing old I got in the last two weeks."

Michener, with the same receding hairline, shot back, "You're still the handsome man I fell in love with."

While Ives was in surgery, Michener paced for hours in the hospital waiting room, stopping only to call home to check on their kids. The idea of losing his spouse after nineteen years and three children was unfathomable. Ives had always been more of the leader of the family, with a commanding presence that had given Michener great comfort. Now their entire world had been upended, their plans to retire to the beaches of Delaware threatened by the absurdity of a bad paper cut.

Finally, a doctor had news. Two teams of surgeons were trying to reconstruct Ives's heart, which was riddled with bacteria, but the odds weren't good. Twice he had nearly died on the table. The doctors could induce a coma to allow his heart to rest before more surgery, but it was a risky procedure of last resort.

"Absolutely. Do it," Michener insisted. "I want him to live."

Somewhere in all their paperwork, they had both made clear they didn't want extensive, life-saving provisions, but in that moment, with three children waiting at home, Michener could think only about saving his husband. In intensive care after eighteen hours of surgery, doctors gave Ives emergency blood transfusions and dialysis for his kidneys. But on the fourth day, the cardiologist called.

"There's been no improvement," the doctor said over the phone while Michener was running their son Jack to the pet store for tropical fish. "Nothing is coming up right."

Michener turned to Jack, who had just celebrated his eleventh birthday. "I'm going to be honest with you," he said, fighting to keep his voice calm. "I don't think Daddy Bill's coming home."

Ives stopped breathing at three A.M. the next morning, just after Michener crawled beside him in the hospital bed. Michener whispered in his husband's ear, "I understand. We had a great life, and everyone loves you."

The nurse knelt down, listened for a heartbeat, and shook her head. Michener draped his arms across his husband's body. "I'll see you on the other side," he said.

Three hundred people came to the funeral, but Michener barely saw their faces. More than anything, he wanted to sit in a quiet room somewhere and grieve, but his children were grieving, too, and for now, that would come first. When the funeral director motioned toward an empty corner of the room, Michener walked over, distracted.

"There's a problem with the death certificate," she said.

Ives had wanted to be cremated, but without a death certificate that listed Michener as the surviving spouse, there was no one to authorize the cremation. Ives's body was in legal limbo. The thought sickened Michener, who went straight home, closed the door to the office so his children wouldn't see, and put his head in his hands.

Later that day, an administrator with the city's health department called attorney Scott Knox about the death certificate. Knox called Al, and two days later, Al went to court to ask Judge Black about expanding the ruling to include Bill Ives. The judge agreed.

With a court order, Michener went straight to the funeral home, terrified that the state would try to appeal and block the cremation. As he watched the casket creep toward the furnace, the whole situation seemed ludicrous and surreal, his husband's sudden death, the delay over the death certificate after years of careful planning, the federal court order. Michener looked around the room, half expecting his husband to jump out from behind the door so they could pick up life where they had left off, an intact family of five.

A month later in September, Michener decided to run a five-kilometer race in his husband's name at the Wright-Patterson Air Force Base in Dayton. Just before the race began, his cell phone rang.

"You have a choice," Al Gerhardstein said. "You can go on and live your life or you can join this lawsuit."

Michener had no interest in any lawsuit, but he thought of Daddy Bill. "Keep me on it," he told Al.

And then he ran and ran.

In the gloomy last days of fall, John Arthur lay dying. The finality of it was everywhere, in the wheezing sounds that John made when he breathed, in the humming of the machines secured to the hospital bed. Paulette was there nearly every day, wanting to stroke John's head or straighten his glasses but worried that even the gentlest movement would aggravate her nephew's nerve endings. On a visit in early October, she pointed to one of the monitors by the bed. It read 100.2.

"Does he have high blood pressure?" she asked Jim.

Jim shook his head. "No, that's his weight now."

A few days later, she sat with John again and whispered, "I love you." John's eyes were fixed on her face, and though his lips moved slowly, he couldn't get out any words. Paulette said, "I know."

John's cousin Keith came by to watch game shows until John drifted off. Curtis Arthur drove in from Toronto, cracking jokes to make his brother smile and waving Jim out of the room to give him a break. A few months earlier, just before John had lost the ability to talk, he looked over at Curtis and said, "It's going to be okay."

"Yeah, I know," Curtis said simply, and left it at that.

On October 21, when President Barack Obama was talking about the Affordable Care Act and the Boston Red Sox were preparing to play the St. Louis Cardinals in the World Series, Jim sat by John's bedside, reading from chapter eight of John's favorite Clive Barker fantasy novel about the secret existence of a parallel world in Liverpool, England. *Immacolata felt a twinge of trepidation. Not because of the man who emerged from the shelter of the column, but because of the company he kept. They moved in the shadows behind him, their panting flanks silken. Lions! He'd come with lions.*

Jim was exhausted and took a break to use the bathroom. He had requested family medical leave to be with John full-time, believing that now that he was married, he qualified for the benefit, which would guarantee his job when he returned to work. But his supervisor had denied the request. "Ohio doesn't recognize your marriage," he told Jim.

"What about professional leave?" Jim asked. "Is my job guaranteed then?"

"No. But it's our intent that you'll have a position to come back to."

"So you're telling me that you're not guaranteeing my job?" Jim said.

There would be no guarantees. The lack of protection was unsettling but Jim left anyway, drawing on their savings to pay the bills.

From behind the bathroom door, Jim heard John's bell ring. Though he was in the final stages of ALS, John could still twist his head slightly to the right, triggering a sensor on the pillow to call for help.

"*What?*" Jim shouted the word, a single moment of exasperation that he instantly regretted. But he was in a world of pain, utterly drained and terrified of what would come next, what death would look and feel and smell like, and how the world would change when John was gone. Jim rushed back into the bedroom. John was mouthing words, but for the first time since his rapid decline to paralysis, Jim couldn't understand.

Jim picked up the phone and called Curtis. "I'm failing him," he sobbed. "I don't know what he needs."

Jim called hospice. "Please, please send someone over now."

Jim went back to John, watching his mouth and tongue rise and fall, as if in slow motion. His eyes were desperate. *In-hal-er.* Jim seized the asthma inhaler on the nightstand and pressed it to John's mouth. Breathe, goddammit, Jim thought. Slowly, John's shoulders relaxed and he closed his eyes. Jim leaned over, his heart pounding, and whispered, "I'm sorry for being an ass."

The hospice nurse came in a few minutes later. "He looks peaceful," she told Jim, and urged him to get some rest. Jim brushed his hand along the side of John's face. "Please wake me if something changes," he said quietly.

In the guest room, Jim tossed and turned in the darkness and wasn't surprised when he heard a soft knock on the door long after midnight. "It's the end," the nurse said.

ALS had been a savage disease, but death came peacefully to John, and for that, Jim was grateful. John's breathing became shallow, and then it simply stopped. The nurse pulled the oxygen tube from John's nose and looked at Jim. "He's gone now."

Jim called Curtis, Adrienne Cowden, Meb Wolfe. Only two and

a half years ago, Jim thought in a stupor, he had made the same series of calls to tell their family and friends that John had been diagnosed with a terminal illness. Sometime in the night, funeral home technicians came in with a body bag, but Jim couldn't stand to watch. Already their condominium felt vast and empty, with every corner crammed with a memory, a painting they had bought together, the table where they'd eaten, the couch they had shared.

Friends came over through the night and into the early morning, huddling in the kitchen and living room. Meb Wolfe embraced Jim, who whispered in her ear, "I don't know how to go on. I don't know what to do."

"John would not want you to stop living," she said.

When Paulette arrived just before noon, she found Jim sitting alone in a chair by John's empty hospital bed. He had sat in that chair every day and night for months, helping John eat or move, playing music, changing the bedpan. Jim looked small and confused, with hunched shoulders and tired eyes.

Paulette fell back against her husband. "Oh my God," she said, "Jim's still in there."

Barbara Cook, who had first told Al about John and Jim, e-mailed early that morning. "Al, you may have already heard this, but I learned this A.M. that John died."

Al responded, "Jim let me know. Thanks for connecting us. They are a beautiful couple."

"They are indeed," Cook wrote. "I told Jim last week that you loved them both."

Al got to work. He e-mailed Aaron Herzig, the lawyer for the City of Cincinnati. "Please be advised that plaintiff John Arthur has died. He was a kind and loving man. Plaintiff James Obergefell . . . will be seeking a death certificate consistent with the orders of the court."

He e-mailed the funeral director, Robert Grunn: "Please apply for John's death certificate consistent with the attached court order listing John as 'married' and James as his 'surviving spouse.' I have alerted the defendants that you will be doing so. Thank you for your service to this loving couple."

Al updated the law firm's website. *In Memoriam: John Arthur. Even as John faced his last days, he was fighting for the rights of all same-sex couples. Part of John's legacy will be the difference he has already made in the struggle for marriage equality. Thank you John for your courage and love.*

Word of John's death spread quickly. The *Cincinnati Enquirer* posted on its website, "John Arthur, who challenged same-sex marriage ban, has died." The Associated Press wrote a story, which was picked up by the Huffington Post and Yahoo! News. Friends shared pictures: ten-year-old John on his way to a rafting trip with his cousin, Keith; forty-one-year-old John drinking a beer in Vancouver for Jim's birthday.

Jim put up his own Facebook message two days later: "On December 31, 1992, I made the best decision of my life—going to a party at John Arthur's house because his roommate invited me. I never left. 20 years, 9 months and 22 days wasn't nearly enough time with the man who taught me what it means to love, laugh, be generous, show kindness, appreciate every day and find the fun in everything that happens. He left the world a better place than he found it."

Staying with friends during those first few days, Jim busied himself with the business of death. Robert Grunn, the funeral home director, called to ask about John's death certificate. "How many do you want?"

Judge Black's court order in July had only been temporary, and a full-blown hearing was planned for December. Jim was terrified that Ohio might change the death certificate if Judge Black or some higher court ultimately ruled in the state's favor, so he blurted out, "I want twenty—or as many as I can get."

Jim went to the funeral home to pick up John's ashes and then contacted a jeweler about fusing his wedding ring with John's. Jim would put some of John's ashes inside a small channel cut inside the ring, and one day he would release the rest into the Gulf Stream. John had asked Jim to release his mother's ashes at the same time so that, in death, they could travel the world together.

John never wanted a funeral, so Jim planned a party. At a local art gallery, he served John's favorite fried chicken and donuts and a menu of John's drinks—Collegiate John (gin and tonic) and Grown Up John (bourbon Manhattan). He put a cardboard box with John's ashes on the bar, and on top stuck a green felt hat that they had picked up years earlier at a Christmas market in Budapest. A friend also put a name tag on the box that read HI, MY NAME IS JOHN.

But after nearly twenty-one years, everything familiar suddenly seemed strange—the tables they had frequented in their favorite restaurants, the grocery store up the street, even their condominium. Jim left town. He would travel to Germany, Iceland, Los Angeles, Wisconsin, New York, Oregon, Anchorage, and Illinois, where he would try to lose himself in a neighborhood of homes designed by Frank Lloyd Wright. But everywhere he went, John came with him, which is why the last thing Jim said at night before he drifted into a dark, restless slumber was, "Good night, John."

Judge Black's groundbreaking ruling in July had made national news, but the decision had only affected the death certificates of two men: John Arthur and Bill Ives. Al wanted to secure accurate death certificates for the rest of the gay community in Ohio, broadening the case and setting significant legal precedent. To do that, he needed Robert Grunn.

The funeral home director with spiked salt-and-pepper hair

and pointy brown glasses was a well-known fixture in Cincinnati's gay community, hosting funerals in a building in the city's business district that had once housed a gay-friendly bar. For a funeral home, it was a pretty place, filled with red leather couches and local artwork, samples of urns discreetly displayed on wooden bookshelves along the brick walls. Grunn had cared for the dead for twenty-five years and hated telling gay clients that they couldn't be named on their partners' death certificates or even included in funeral plans if the parents of the deceased objected.

If Grunn became Al's third plaintiff, it would clear the way for a ruling that allowed all funeral home directors in Ohio to issue death certificates acknowledging out-of-state marriages among gay couples, giving spouses the right, among other things, to make burial and estate decisions.

But the state vigorously objected to any expansion of the case, and eight days after John Arthur's death, Judge Black called for a hearing to sort through the matter.

"Good afternoon, Judge. Al Gerhardstein for the plaintiff," Al said in the same courtroom where he had initially presented the case. "And with me is James Obergefell and Robert Grunn, who are plaintiffs."

"I would say this to anyone in any civil case," Judge Black said a moment later, looking directly at Jim. "Mr. Obergefell, on behalf of the Court and the community, I express my condolences upon your loss."

Jim blinked back tears and mouthed, "Thank you."

Before the hearing, Al had submitted a legal brief arguing that funeral home directors, who faced criminal penalties for making false statements on death certificates, needed clarification on whether to follow Ohio's marriage recognition ban or Judge Black's recent orders concerning John Arthur and Bill Ives.

The state wanted Grunn dismissed as a plaintiff, arguing that he didn't have a direct tie to the case and hadn't suffered any harm. Assistant Attorney General Bridget Coontz stood up and told the judge, "Mr. Grunn does not allege that he has a close relationship with anyone whose rights he's trying to vindicate. . . ."

"Who better to raise this issue as to the death certificate of the same-sex couple than a funeral home director who services, in large part, the gay community?" Judge Black asked.

"The individual plaintiffs who are applying for the death certificate," Coontz replied.

"So you do it one by one?" the judge asked. "As people die, they rush into court within twenty-four hours?"

The hearing ended quickly, and two days later, Judge Black issued another ruling, his most crucial one yet. "As fully anticipated by all parties," Judge Black wrote, "Plaintiff John Arthur died very recently on October 22, 2013. The question now arises whether this lawsuit dies with him."

The answer, the judge decided, was no. Al would get to keep Grunn as a plaintiff, broadening the case against the State of Ohio. At a final hearing in December, Judge Black would decide whether the marriages of gay couples across Ohio deserved recognition in death, just as he did for Bill Ives and John Arthur.

14

TWENTY-ONE YEARS FROM MIDNIGHT

AL LOVED to sue people. It was the surest way to make the law work for the weak, the most satisfying kind of advocacy. On the best days, he could find legal doctrine to combat the misfortunes of his clients, and the worst days weren't particularly awful because lawsuits often brought attention to problems with no legal fix. Losing in court could set bad precedent, but early in his career, Al had decided to use the law with purpose, no matter the outcome, as long as the case moved to solve a significant problem.

His lawsuit against Ohio had produced some early wins, but as he scrambled to prepare for the December hearing, Al braced for uncertainty. If he won in Judge Black's trial court, the state would likely take the fight to the Sixth Circuit Court of Appeals, where three of more than twenty judges would be randomly selected to decide the case. Al knew that even sympathetic judges could find him on the right side of fairness but the wrong side of law, and ultimately rule against him.

A few months earlier, Al had heard from the American Civil

Liberties Union (ACLU) in New York, where lawyers worried that a fresh loss in the Sixth Circuit would set damaging precedent. Promising freedom-to-marry cases were advancing in Virginia, Pennsylvania, and North Carolina, and appeals from those states would be heard by federal courts considered more liberal than the Sixth Circuit. A marriage case from Nevada was already before a federal appeals court considered the most progressive in the country. The ACLU wanted good case law and feared that a defeat in the Sixth Circuit before rulings in those other states could set back the marriage movement.

Al was torn. He wanted to support national strategy, which he knew was highly coordinated and backed by all of the country's major gay rights groups. In 2005, a year after voters in Ohio and other states had banned same-sex marriage, leaders of the marriage movement gathered in a hotel in New Jersey and penned a call to action for an eventual victory at the U.S. Supreme Court. They would choose battlegrounds carefully, build strong legislative campaigns, and try to dramatically change the way people viewed gay couples and their families. Then, only with a critical mass of states and strong public support, they would bring a case to the Supreme Court. A nonprofit called Freedom to Marry, led by lawyer Evan Wolfson, who had helped win the first same-sex marriage victory in Hawaii, would devote resources to state organization and public education.

In just a handful of years, the campaign had made enormous progress through bills in state legislatures, ballot-measure victories, and lawsuits led by Lambda Legal, the ACLU, the Gay & Lesbian Advocates & Defenders, and others. By the end of 2013, eighteen states plus the District of Columbia would allow same-sex couples to marry.

But Ohio, assigned to the Sixth Circuit Court of Appeals, was considered among the most difficult to turn.

Al weighed his options and ultimately decided that he had clients with an immediate need and a Supreme Court decision in *Windsor* that seemed to directly support his case. He couldn't wait on the national movement and he couldn't worry about losing. "If precedent means anything," he told the ACLU's James Esseks, a Harvard-educated attorney who had worked on the marriage issue for more than a decade, "we should be able to follow a Supreme Court ruling."

Esseks ultimately agreed. Al's lawsuit was the first in the country narrowly focused on the recognition of gay marriage rather than the right to marry itself. Esseks, who had been cocounsel for Edie Windsor at the Supreme Court, decided that if there was any fact pattern that could draw a favorable ruling from the Sixth Circuit after the *Windsor* decision, this was it. The ACLU helped Al prepare for the hearings before Judge Black, and Esseks eventually signed on as Al's cocounsel.

Later, Al described his decision to push forward to Jim Obergefell. "I have clients that are being harmed. Of course it's the right thing to do."

On a Wednesday morning eight days before Christmas 2013, Al put on a red tie with pictures of snowmen, walked one block to the federal courthouse, caught the elevator to the eighth floor, and prepared to take on Ohio's marriage recognition ban again. Jim Obergefell, David Michener, and Al's newest plaintiff, funeral home director Robert Grunn, sat beside him at the plaintiff's table.

Judge Black walked in through a side door at ten A.M. and said, "The court's ready to proceed. On behalf of the plaintiffs, Mr. Gerhardstein."

Al stood up. "May it please the court, this case is about love surviving death, whether the plaintiffs, who were married in other states, can require Ohio to recognize their valid, legal, same-sex marriages on their Ohio death certificates. We need to remind ourselves where we've been. When James Obergefell and John Arthur started this case, they really had a simple goal. John Arthur was going to die and they wanted his Ohio death certificate to accurately report that he was married and that James Obergefell was his surviving spouse."

Jim shifted in his chair behind Al, momentarily surprised to hear his name in the courtroom. It had been a lonely eight weeks since John's death, and Jim had just returned from a weekend trip in New York City. He had found the street corner where John, two months after the 2001 terrorist attacks, wandered past Mayor Rudy Giuliani and his security detail, too busy admiring the local architecture to notice. "Do you have any idea who that was?" Jim had said.

Jim remembered that day in bits and pieces as he roamed alone along frigid city blocks. *What do I do now? Where do I go from here?* Grief was a stubborn, punishing companion, and Jim had walked blindly until dusk, when it was too cold to walk anymore.

As the hearing progressed, Judge Black stopped Al. "The state makes a big deal about the people of Ohio [who] voted to establish these bans. The law is clear that, you know, voters can't pass a statute to violate the Constitution. Is that where we are?"

"Right. Yes," Al replied.

" . . . Well, just big picture, I suppose there were statutes passed by voters that refused to permit marriages between people of different races," the judge said.

"Yes . . . "

"So your position is, with all due respect, the mere fact that the majority of the voters in 2004 voted for something, if it is unlawful discrimination, the court is required to act?"

" . . . That's where the courts come in," Al said swiftly. "The courts are here to say, 'All right, we have these fundamental principles that we're going to honor, equal protection, due process, and if the majority passes a law that violates those fundamental principles, it's the job of the court to strike it down. Strike it down wide? No. Strike it down as narrowly as possible so that the democratic process could fix it. And that's all we're doing here."

"Just fix the death certificate issue?"

"Well, the democratic process may choose to fix more than that, but the only issue before this court is the failure to recognize marriages when citizens come before a registrar seeking a death certificate that has the marriage on it—that should have the marriage on it."

Al paused and lowered his voice to emphasize his next point. "I mean, yesterday was my anniversary, forty-one years married. If I die tomorrow, I am sure that my wife would want on my death certificate, and I would, too, the fact that the biggest event in my life is properly noted. And that's all same-sex marriage, death certificate, remedial relief will say. Yes, that was the biggest event in John and James's life. That was a defining aspect of themselves as a couple. And they deserve to have the same dignity that my wife and I would have."

Assistant Attorney General Bridget Coontz could sense the strength of the opposition as she watched Al Gerhardstein, gesturing angrily with his hands, and Jim Obergefell, grim-faced and pensive. "May it please the court," she said, rising, "counsel for plaintiffs have made very clear that this case is about recognition of out-of-state, same-sex marriages in a very narrow context, death certificates. That's it. And accepting as a given that the plaintiffs are not challenging Ohio's right to define marriage as between a man and a woman, this case is very narrow. So even if the plaintiffs get all of the relief that they seek, same-sex marriages will still not be permitted in Ohio and it won't be recognized in Ohio anywhere other than in the plaintiffs' death certificates and the death certificates that are issued by Mr. Grunn."

Coontz pointed to the specifics of the *Windsor* ruling, which required the federal government to recognize same-sex marriages but did not directly address whether the states had to recognize them, too.

"The whole subject of domestic relations belongs to the laws of the state," she argued. "And this is not a new concept.. . . The Supreme Court has never said that one state has to recognize same-sex marriages performed by another. It certainly didn't say that in *Windsor*."

"But you do acknowledge that voters can't pass unconstitutional stuff, right?" Judge Black asked a few minutes later.

"Yes," she said. "Yes, I do."

Jim listened, annoyed. Before they were married, Jim and John had celebrated their anniversary on New Year's Eve. It would have been twenty-one years that month since their chance encounter just before midnight at the Abbey, when Jim, feeling bold and free, had sidled up behind John in the kitchen. Jim thought about John's death certificate, a single, crucial piece of paper that officially recognized their marital status, stashed safely in the nightstand by his bed.

Coontz went on. "The terms *husband*, *wife*, and *spouse* collectively appear hundreds of times throughout Ohio's Revised Code, so changing who is and who is not married under Ohio law has sweeping consequences. The voters' desire to evaluate and make that change deliberatively, if it's going to happen, is entirely rational. . . ."

"The need to act cautiously has never been a valid defense to stopping unconstitutional activity, has it?" Judge Black asked.

"No, but that presumes that the statute is unconstitutional, and the state's position is that it's not."

"Very well."

". . . Ohio's desire to retain the right to define marriage is rational. Ohio doesn't want Delaware or Maryland to define who is married under Ohio law. To allow that to happen would allow one state to set the marriage policy for all others."

Again, Judge Black intervened. "This court is not going to have anything to do with ordering Ohio to perform same-sex marriages. That's not before the court. It is a very limited issue. It is when one state gives you something and you come to Ohio, can Ohio take it away without due process?"

"It's recognition," Coontz said.

"Right."

"It is. But what the plaintiffs are asking is that Ohio treat their marriages as valid in Ohio when Ohio law provides that they're not. . . . This case illustrates that changing laws through the judicial process in the piecemeal fashion that plaintiffs seek can create inconsistencies through Ohio law. And it is rational for Ohioans to say, 'No. We want to do this through the democratic, not through the judicial, process.' "

Judge Black pressed. "I mean, you can't say that we've got a rational state interest in this because we want to control it and we want to pass stuff that doesn't recognize the role of the United States Constitution. . . . Why [can] Ohio, when faced with a bunch of coequal states, decide that a marriage license from Delaware is worthless . . . ?"

"Because Ohio has the right to define its marriage policy . . . Ohio is not invalidating someone else's marriage. We're simply saying that, for Ohio's purposes, their marriage is not recognized under the law of the State of Ohio. Congress has said that we're allowed to do that, and that's the posture of this case."

Judge Black paused. "Well, I suppose I better recognize it on the record. You're a good lawyer."

"Thank you, Judge," Coontz said, and sat back down in her seat.

"Very well," the judge said. "Well, oral argument has been helpful to the court. It's been helpful all along. Briefing is excellent, and the court's preparing to act."

"All rise," the deputy called, and Judge Black left the courtroom.

Al picked up his worn satchel crammed with six months of legal documents and walked outside with Jim and the other plaintiffs. Television crews were waiting on the steps of the nine-story courthouse in the heart of the city's business district.

Jim turned to the cameras. "I'm hopeful, nervous. But I'm doing the right thing for John and me, for David and Bill, and for every other gay couple in this state."

On December 23, five days after the hearing, Judge Black released his forty-eight-page final decision. "Once you get married lawfully in one state, another state cannot summarily take your marriage away. . . . Under the Constitution of the United States, Ohio must recognize on Ohio death certificates valid same-sex marriages from other states."

State officials announced they would appeal.

From his office across the street from the federal courthouse, Al pored over every line of the lengthy opinion ordering Ohio to permanently change the way gay people were described in death. The judge had focused heavily on the history of discrimination against gays and the campaign behind the 2004 same-sex marriage ban in Ohio, when Citizens for Community Values had warned the state's employers that "sexual relationships between members of the same sex expose gays, lesbians and bisexuals to extreme risks of sexually transmitted diseases, physical injuries, mental disorders and even a shortened lifespan."

But it was a single footnote that caught Al's eye. On the bottom of page 7, the judge cited a legal brief that he had received from Citizens for Community Values. "Among its many remarkable and fundamentally baseless arguments," the judge wrote, "one of the most offensive is that adopted children are less emotionally healthy than children raised by birth parents."

Al read the footnote once, twice. He started thinking about what needed to come next, how to make the law work for those less fortunate.

Home alone in his condominium, Jim posted on Facebook: "We won!!!"

But it was John's victory, too, and he would never know how his death had inspired the beginnings of change in Ohio. Over New Year's Eve, on a trip to see the two foreign exchange students that Jim and John had hosted in Cincinnati years earlier, Jim posted another message: "You aren't with me in person, John Arthur, but I'm still celebrating. Twenty-one years ago today, you entered my life and made it bigger, brighter and happier than I deserved or ever expected. I love and miss you."

Then Jim said good-bye to 2013.

Two new lawyers had watched Al closely during the December hearing, furiously taking notes on legal pads from their seats near the back of the courtroom. Early that morning, Shannon Fauver and Dawn Elliott left their Louisville legal practice, in a ramshackle Victorian house near two bars and a set of train tracks, to make the ninety-minute trek to Cincinnati in Fauver's Volkswagen.

Fauver's family had settled in Kentucky in 1876, and since her great-grandfather, grandfather, and parents had been lawyers, it seemed only fitting that Fauver set up her own practice after a stint in the Peace Corps. In 2006, she was the only lawyer in town to advertise in a telephone directory geared toward the gay community and began to attract a steady stream of gay clients needing help with bankruptcy or Social Security matters.

Fauver met Elliott in traffic court when Elliott was pushing to secure a plea deal for a drunk driver. Born on the city's less affluent west side, Elliott paid her own way through law school at the University of Louisville, eventually becoming an expert in family law. She agreed to join Fauver's practice. The two lawyers made a dynamic pair, juggling court dates and families and taking a keen interest in local politicians, who set up campaign signs on the firm's front lawn.

When the *Windsor* decision was released, they poured two glasses of bourbon and read through the ruling late into the night, certain that the Supreme Court had just given them an opening to file a lawsuit on behalf of Kentucky's gay couples. Through friends, they met Greg Bourke and Michael De Leon, who had traveled to Niagara Falls to marry in 2004 after twenty-two years together. They had two adopted children, but Kentucky would only recognize one father as the adoptive parent. Fauver and Elliott started drafting a complaint that challenged Kentucky's marriage recognition ban. They drove to Cincinnati for the first time in July 2013 to watch Al argue in federal court for a temporary order that required the state to list John Arthur as a married man on his death certificate. Two weeks later, Fauver made a cold call to Al's office, hoping he would review their complaint in Kentucky. "Could you let me know how it looks?" she asked.

"You're filing tomorrow?" Al was surprised. "Have you ever done this before?"

"No," Fauver admitted. "But we're doing whatever we need to do for our clients."

"You sure you want to file tomorrow?"

"We're filing in the morning."

"If you need anything," Al said after he briefed Fauver on the fundamentals of civil rights cases, "you let me know."

Through the fall and early winter, Al coached the two lawyers over the phone as they waited on word from Kentucky federal judge John G. Heyburn II. Al described case law, the history of gay rights, and Issue 3 in Cincinnati.

"I want to be like you when I grow up," Fauver, who was forty-four, told Al one afternoon.

Al chuckled.

"No," Fauver said. "Really."

Sitting in the back of the courtroom in Cincinnati in December, Fauver and Elliott wondered whether, after all their calls, Al would somehow spot them in the spectators' section, Fauver with pale skin and red hair, and Elliott, African American and four years younger, with a wide smile and a head of curls.

They introduced themselves after the hearing, and on the drive home, they talked about Al's lively performance in court, the way he waved his arms when he was making a particularly critical point, the way he personalized his arguments by citing his own marriage. "If we have oral arguments," Fauver told her law partner, "you're going to have to do it."

With their finances stretched thin and compensation from a potential win in court months or years away, the two lawyers had asked a larger Louisville firm to help with the case. Civil rights lawyers Dan Canon, Laura Landenwich, and L. Joe Dunman signed on.

Early on, Fauver and Elliott had also reached out to the ACLU in Kentucky, but the organization worried about a loss in the Sixth Circuit.

"Screw it," Fauver said. "We're going to file anyway."

"Pay it back," Elliott added, nodding. She was an African American woman who had married a white man, a union that wasn't legal until the Supreme Court in 1967 ended all race-based restrictions on marriage. Somebody fought for me, Elliott decided. And now it's my turn.

More plaintiffs were added to the case, and on February 12, 2014—two months after Judge Black's ruling in Cincinnati—

Judge Heyburn issued his ruling in Kentucky: "Our Constitution was designed both to protect religious beliefs and prevent unlawful government discrimination based upon them. . . . In the end, the Court concludes that Kentucky's denial of recognition for valid, same-sex marriages violates the United States Constitution."

Kentucky attorney general Jack Conway decided against an appeal, telling *Time* magazine, "I know where history is going on this and I know what was in my heart. . . . I draw the line at discrimination." But Kentucky governor Steve Beshear hired outside counsel to appeal on the state's behalf.

Fauver, Elliott, and their cocounsels, who had won the country's second major ruling on marriage recognition, would follow Al to the Sixth Circuit Court of Appeals.

15

SEVENTY-TWO HOURS

COOPER J. TALMAS-VITALE was born in Ohio just after dinnertime on April 19, 2013, six pounds, eleven ounces, with a wisp of brown hair and the tiniest fingers and toes his fathers had ever seen. The social workers had told them to wait three days before falling in love, just in case the baby's birth mother changed her mind and called off the adoption. But Joe Vitale and Rob Talmas had spent six months dealing with the paperwork to become parents, one year trying to find a birth mother, ten hours racing from their home in New York City to the hospital delivery room in Ohio, and forty-five minutes on a frantic buying spree at a baby superstore that yielded a stroller, a portable crib, and a stuffed Curious George, wearing a red T-shirt and a big grin.

"Don't buy too much. It may not happen," Vitale warned his husband, who was stuffing the shopping cart with diapers. The social worker had been adamant: three days.

But Vitale was in the delivery room when Cooper was born. He texted a photo to his husband, who was praying in the hospital chapel. It was a picture of the blue-eyed newborn, swaddled in a white knit cap, the first of 6,500 photos they would capture of their son in the coming months. He's ours, Talmas thought as he sprinted to the nursery, and the three days flew right out the window.

In the Catholic hospital on the outskirts of Cincinnati, the nurses embraced the two gay men from New York City. One helped Vitale scrub in and cut the baby's umbilical cord. Another offered the empty room across the hall so the new fathers could take part in nighttime feedings. A third put Talmas in a wheelchair, placed the sleeping infant in his arms, and rolled them out of the hospital to an SUV stuffed with so much baby gear, they had trouble closing the trunk.

Talmas e-mailed his sister and a small circle of friends, dubbed Team Cooper, who had helped with the adoption: DONE and DONE! Cooper is ours.

He e-mailed a picture to his parents, who hadn't been told about the baby in case something went wrong: Here is your grandson.

Talmas's father, a retired economist, called minutes later, thinking the e-mail was a joke. Then he grew quiet.

"Wait. Is this really our grandkid?"

"It's your grandkid," Talmas said.

They drove straight back to New York City, blinking every few minutes at their snoozing newborn, wrapped up like a burrito in the backseat. For sixteen years, it had just been Joe and Rob, along with a boisterous brood of brothers, sisters, nieces, and nephews scattered across the suburbs of New York and New Jersey. With careers in insurance and human resources, they lived in a one-bedroom condominium on New York City's East Side and traveled to Rhode Island on the weekends, where they passed the time fishing or cooking dinner for friends.

But there had never been any question of one day starting a family. They married in lower Manhattan's city hall in 2011, three months after Governor Andrew Cuomo signed the New York Marriage Equality Act, and started placing calls about private adoption.

And now Cooper was home, in a nursery with a toy tepee, and it seemed as if it had always been this way, with baby gates and sticky floors and toy trucks perched on the window ledge. Cooper holding a bottle. Cooper eating his first bite of spaghetti. Cooper in snowman footie pajamas. Cooper at the pediatrician with hiccups that his fathers took for convulsions, much to the amusement of the office staff. They called their son "Little Prince."

Cooper was just learning to talk in January 2014 when Vitale got a call from the adoption agency. The nonprofit organization in western New York had never had exclusionary policies against gay parents, and Vitale and Talmas, both in their mid-forties, had spent months working with Adoption STAR's social workers and parent mentors.

"We have a problem here, bad news," associate director Michael Hill said.

"What bad news?" Vitale tensed and thought of Cooper, who had started toddling to the front door, drooling with outstretched arms, every time they came home from work. It was the most sublime five minutes of the day.

"Ohio will only put one of your names on Cooper's birth certificate," Hill said. "There's only space for one father—not two."

"You've got to be kidding me." A surrogate court in New York had already finalized Cooper's adoption and had listed both men's names on the order. All that was left was the birth certificate from Ohio, but Vitale knew that it was the single most important legal document, needed for Cooper's schooling, travel, health insurance, and medical care. How could they be forced to decide which father to list and which father to leave off? And how could he break the news to his husband, who had been adopted at birth by a Jewish family and had never known his biological parents? Talmas cherished the idea of a strong and uniform family identity. "Totally unacceptable," Vitale seethed.

"Listen," Hill said. "There is a death certificate case going on in Ohio with this guy named Jim Obergefell. You should call his lawyer on Monday."

It was Sunday. "I'm calling now," Vitale said, and hung up.

He dialed Al Gerhardstein's law office, pressed Al's extension, and prepared to leave an urgent message. But Al picked up on the first ring.

"My name is Joseph Vitale. My husband is Rob, and our son is Cooper."

"You're the New York couple," Al replied easily. "And you want your son to have a correct birth certificate."

"It's not about Rob and me," Vitale said, somewhat mollified by Al's quick response. "It's about Cooper."

From his office in Cincinnati, Al had been looking for parents like Vitale and Talmas for weeks, ever since he decided that the next step—now that Judge Black had ruled on what happened at the end of life—would focus on life's beginnings. From birth to death, Al thought.

He started calling family attorneys and adoption agencies and quickly discovered that legally married same-sex couples that adopted babies in Ohio or had them through artificial insemination were denied birth certificates listing the names of both parents. There was legal backing to do it, since state law denied recognition to married gay couples.

To Al, it was another practical problem that he could bring to court, much like the case on death certificates. He knew that birth certificates gave parents the legal authority to approve medical care, apply for health insurance, travel with their children, and interact with schools and childcare workers. Yet Ohio was forcing same-sex couples to choose one parent over the other on an official document that established a child's identity and family.

Al thought about his grandson, Oliver, who was about to celebrate his second birthday. Before bedtime, he would fix his gaze on Al, not particularly focused on the words in Al's bedtime story about a wise owl that would save Oliver and his friends but on the drama in Al's voice.

Al started gathering the names of possible plaintiffs and heard about Vitale and Talmas through their adoption agency. Then, on a Sunday afternoon, Vitale called.

In mid-January 2014, Cooper J. Talmas-Vitale became Adopted Child Doe, the youngest plaintiff in Al's second lawsuit challenging Ohio's ban on marriage recognition.

Al found two more plaintiffs in the suburbs of Kentucky, pregnant with their second son.

The first time around, Nicole Yorksmith had struggled through six rounds of artificial insemination before becoming pregnant. Her wife, Pam, went out again, and again, every night for five straight months to the convenience store around the block to bring thirty-one-year-old Nicole peanut butter milkshakes or Fruit Roll-Ups or the occasional can of SpaghettiOs, whatever satisfied the day's cravings.

Grayden Yorksmith came unexpectedly in 2010, after the women spent a warm night at a pumpkin patch.

"Oh my God, Pam," Nicole said as they were getting ready for bed. "I think I peed on myself. Help me get to the bathroom."

There was a pool of water on their hardwood floor. "Baby, I think your water broke."

"No. I think my bladder is doing something funny."

Pam Yorksmith smiled at her wife, whose brown hair was flying every which way. Six years older with a career in technology,

Yorksmith had always been the voice of reason; her first gift to Nicole after they met at a friend's murder mystery party in 2006 had been a humidifier because dry air made Nicole sniffle.

Yorksmith used pillows to prop her wife up in bed, and when the contractions started a few hours later, they met their obstetrician at a hospital in Cincinnati. Grayden was a plump, happy baby, and the two mothers moved their son and a feisty Labrador-pointer mix named Kahlua from Cincinnati to a new house on a cul-de-sac in northern Kentucky, with a bright-yellow kitchen and a mimosa tree that drew hummingbirds in the fall.

A single piece of paper, Pam Yorksmith decided, shouldn't have mattered much to the busy family of three. She was running a technology consulting company and Nicole was commuting into Cincinnati, where she worked in human resources at Procter & Gamble. But when Grayden's birth certificate arrived in the mail with only her wife's name on it, it seemed to Yorksmith as if she no longer officially mattered, all those late-night feedings and trips to the pediatrician, the crushing tenderness she felt for her son, denied by the State of Ohio. For the first time as a new mother, she felt vulnerable and afraid, and the wills and parenting agreements that she and her wife had drawn up together, just in case, no longer seemed as binding.

The two women had traveled to California to marry in 2008. In the upstairs hallway of their new house, they kept a wall clock permanently set at 6:30 P.M. to commemorate the moment they exchanged wedding vows in white dresses, with Pam in a lavender train. They stopped a second clock at 5:08 P.M. to memorialize Grayden's birth.

But Pam Yorksmith was the non-biological mother, and without standing on an Ohio birth certificate, she worried about losing access to her son. Something could happen to her wife. Something could happen to their marriage. A few of their gay friends had split, and in particularly turbulent breakups, "non-bio moms" had to fight to see their children.

"I do the four A.M. feedings," she lamented, "but I am always less than."

And then Nicole became pregnant with their second son.

They learned about Al Gerhardstein's lawsuit on a Facebook page for gay women with children. In January, with her wife five months pregnant, Pam Yorksmith called Al. "I want a secure legal tie to my children," she said. "I don't want to have to worry every day if something were to happen to Nicole that someone is going to come and take my kids away from me."

"Just like we worked on the end of life," Al said, "we also want to work on the beginning of life. A birth certificate should reflect the true nature of what's happening in your life, whether Ohio says that it's true or not."

Al invited the Yorksmiths to join the lawsuit. "There will be a lot of press around this," he warned.

But there really wasn't much to discuss. Busy preparing a second nursery with blue walls and safari prints, the two mothers signed right up.

Instead of litigating case by case, every birth and every death, Al decided that now he would take on Ohio's entire marriage recognition ban. In the early weeks of 2014, he and his team drew up a lawsuit in the names of Joe Vitale, Rob Talmas, and Adopted Child Doe; Pam and Nicole Yorksmith; and two other female couples, both married and expecting to deliver babies in Ohio.

Al could have challenged the law's main tenet: the blanket ban on same-sex marriage in Ohio. In other states, civil rights lawyers were pursuing marriage equality cases through high-profile lawsuits that appeared destined for intense battles in state legislatures or the courts. But in the topsy-turvy fight for gay rights, Al wanted to keep his arguments closely synced to the Supreme Court ruling in *Windsor*, which required the federal government to recognize married gay couples. He would attack only the portion of Ohio law that dealt with marriage recognition, a single step in a long game that he hoped would set fundamental precedent.

He considered the state's policy on birth certificates a callous disregard for families, and in a legal brief, his team wrote, "Every time these parents and their children, as well as third-parties, look at the children's Ohio birth certificates, they will see official disrespect for the families and legal insecurity for their relationships."

In Al's law office, the staff hung up photos of Cooper and his fathers; Grayden and his mothers; Kelly McCracken and Kelly Noe, who were married in Massachusetts and expecting a baby girl; and Brittani Henry and Brittni "LB" Rogers, who were married in New York and expecting a baby boy. The law could annihilate bad policy. Al thought about that every time he walked past the photos, to his desk or his law library or across the street to the federal courthouse, where a plaque at the front entrance described the Bill of Rights.

Al knew the issue of marriage equality to many Americans often seemed more philosophical than tangible, a fight in someone else's backyard. But reporters and other civil rights lawyers had begun to call, drawn to the struggles of his plaintiffs, particularly John Arthur and Jim Obergefell. *Every civil rights case starts with a story.*

For his new case on birth certificates, Al decided to partner with Susan Sommer, a Yale-educated constitutional law attorney with the national group Lambda Legal. Al would remain lead counsel.

On a clear Monday morning in mid-February, Al put on one of his neckties with pictures of smiling children, walked across the street to the federal courthouse, and sued Ohio for the second time. All of his plaintiffs were in town except for Rob Talmas, who worried about bringing Cooper into Ohio without a birth certificate that listed the names of both fathers.

In his office afterward, Al smiled at Joe Vitale and the three pregnant women and their wives, all due to have babies that summer. Reporters were waiting in his tiny law library, where a conference table had been pushed to the side of the room to make space for a press conference. Vitale, who sold insurance to New York City's top law firms, had been startled when he walked in earlier that day to find frayed green carpeting and battered wooden desks, like something out of a 1970s sitcom. But he felt instantly at ease when Al said just before the press conference, "Talk from the heart. Talk about why this is so important to you."

Facing local television crews, Vitale twisted the wedding ring on his finger. He had never spoken publicly about the right to marry and had certainly never been on the evening news. He looked at Al, talking calmly into the bank of microphones as Grayden Yorksmith played with toy trucks on the floor. "A marriage is a marriage and a family is a family," Al said. "A family is a loving, nurturing group of people, and the identification document when the children come along is the birth certificate, and it ought to be right."

With a shaking voice, Vitale added, "We want to be afforded the same benefits and rights as every other citizen of the United States."

That afternoon, news of the lawsuit spread. "The idea that two men on a birth certificate or two women on a birth certificate, I'm sorry, it just defies logic," Phil Burress, with Citizens for Community Values, told Cincinnati's WLWT News Five. "It's absolute nonsense. The case should be thrown out of court."

Later, Ohio Attorney General Mike DeWine would argue that

what had started as a narrow challenge to Ohio's policy on death certificates had turned into an all-out effort to strike down a critical portion of state law, one that had been approved by a majority of voters and affected more than just birth and death certificates.

Vitale didn't dwell on the opposition. He left Ohio after the press conference, landed in New York just before dinner, hopped into a taxi, and practically ran with his suitcase from the elevator in his high-rise to his front door down the hall. He stumbled inside and kissed his husband and Cooper, who was drinking a last bottle before bed. Vitale was happy. He was home.

Over the years, Jim Obergefell (bottom right), and his longtime partner, John Arthur (top), had talked about getting married, but Ohio's voters banned same-sex marriage in 2004 and the federal government did not recognize state-sanctioned marriages among gay couples. Jim, who lost his mother while he was in college, grew close to John Arthur's mother, Marilyn (bottom left). She was a frequent guest at their house located on a hillside overlooking Cincinnati's Mill Creek Valley. John and Jim nicknamed the property the "Big House" and filled the rooms with work by local artists. (Courtesy of Jim Obergefell)

John Arthur (left) and his younger brother, Curtis, forged a close relationship early on. Years later, after John was diagnosed with ALS, Curtis left his job in Saudi Arabia and moved to Toronto to be closer to his brother. (Courtesy of Paulette Roberts)

John (top center) and Curtis Arthur grew up in a turbulent home, and John fought frequently with their father, Chester (bottom right). When John started dating men, Curtis told Chester, "If you cut John out of your life, you're going to cut me out, too." (Courtesy of Paulette Roberts)

John Arthur's favorite aunt, Paulette Roberts, loved John like her own from the moment he was born. In 1977, she took John (left), Curtis Arthur, her son Keith Cassidy (far right), and friend Nancy Cutler on a river rafting trip in West Virginia. (Courtesy of Paulette Roberts)

Two years after Jim Obergefell and John Arthur started living together, family and friends gathered at Chester Arthur's house, and Jim was struck by the similarities between John and his father. From left to right: John Arthur's uncle Mike Roberts, Curtis Arthur, longtime friend Meb Wolfe, Jim Obergefell, and Chester Arthur. (Courtesy of Paulette Roberts)

By 1995, John Arthur and his father, shown here at a family wedding, had formed an uneasy truce. "Well," Chester Arthur would tell his son after the ALS diagnosis in 2011, "sometimes you've just got to play the hand that life deals you." (Courtesy of Paulette Roberts)

Jim Obergefell and John Arthur fell in love in 1992 at a New Year's Eve party at John's house in Cincinnati, which they called the "Abbey." Within weeks, Jim left graduate school and moved into the house with John. They posed for a picture there in 1993. (Courtesy of Jim Obergefell)

Jim Obergefell was a straight-A student in Sandusky, Ohio, the youngest of six siblings. His father, Arthur, worked in factories and manufacturing plants. His mother, Mary, was a librarian. In 1972, he dressed up with his parents for a rehearsal dinner for his brother's wedding. (Courtesy of Jim Obergefell)

In the early 1990s, Al Gerhardstein had a growing civil rights legal practice in Cincinnati. Al and his wife, Mimi Gingold, took their children Ben (right), Adam (center) and Jessica to the First Unitarian Church, which would later honor Al for his work on gay rights. (Courtesy of the Gerhardstein / Gingold family)

Al Gerhardstein's youngest child, Jessica, loved to go to her father's office. Years later, Al told her, "You know, you might be frustrated if you're on the sidelines and not able to engage directly in the systems that you want to change." Jessica would one day decide to go to law school. (Courtesy of the Gerhardstein / Gingold family)

Jim and John loved to travel; in 2009 they took a Mediterranean cruise (above). Three years earlier they traveled to Sweden (below). (Courtesy of Jim Obergefell)

In the months before John's ALS diagnosis in 2011, Jim and John spent time in New York City, touring museums, seeing shows, and shopping. At home in Cincinnati, they had sold their large house, bought a condominium, and renovated every room. (Courtesy of Jim Obergefell)

In February 2014, Al Gerhardstein sued the State of Ohio on behalf of four gay couples, including Pam and Nicole Yorksmith and Joseph Vitale and Rob Talmas, who wanted to secure accurate birth certificates for their children. At a press conference in Cincinnati, Al said, "A family is a loving, nurturing group of people" and their children's birth certificates "ought to be right." (AP Images / Al Behrman)

Al Gerhardstein (left), his law office manager Mary Armor (center) and his law partner Robert Laufman (right) at the annual Stonewall Cincinnati dinner in 2000. They spent years working together in a civil rights practice and were unafraid of taking on the most powerful local institutions, including the City of Cincinnati. (Courtesy of Mary Armor)

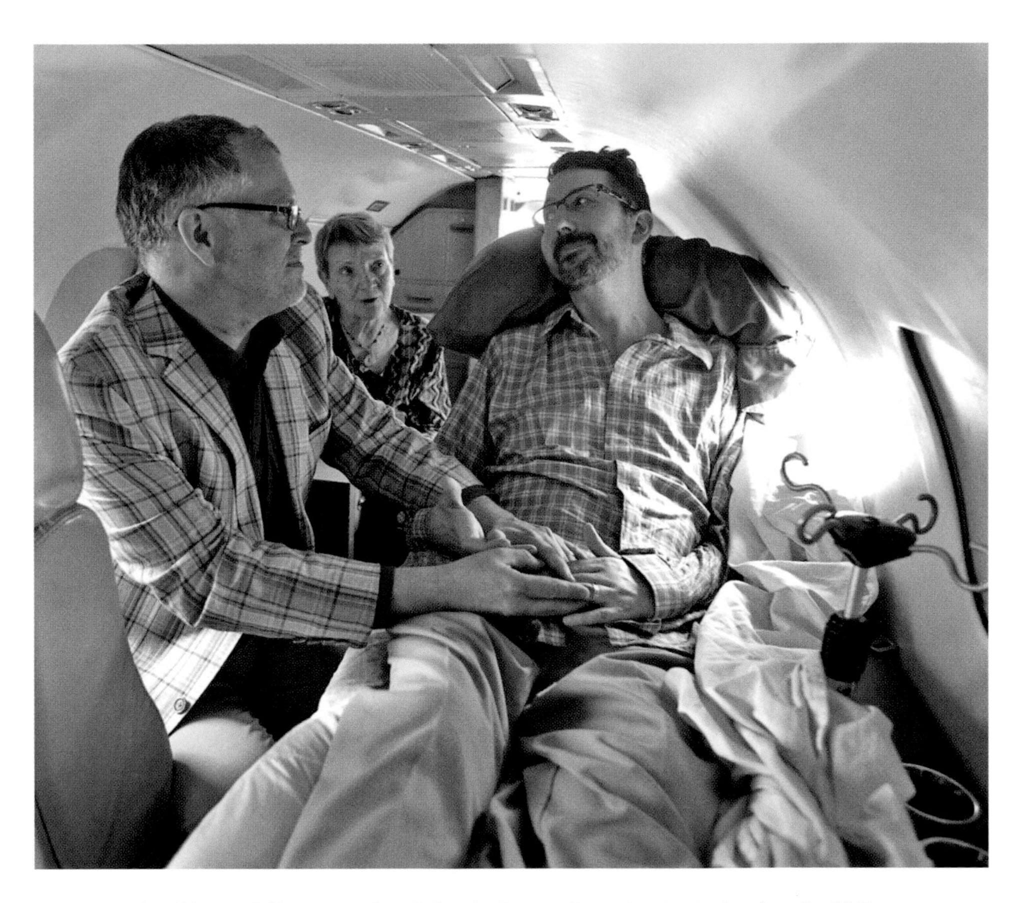

Jim Obergefell proposed to John Arthur on June 26, 2013, the day the U.S. Supreme Court issued an historic ruling that required the federal government to recognize state-sanctioned marriages among gay couples. Jim chartered a medical plane and flew with John to Maryland, where same-sex marriage was legal. John's aunt Paulette Roberts (center) performed their marriage on a tarmac outside of Baltimore. (From The Cincinnati Enquirer, 2015-06-26)

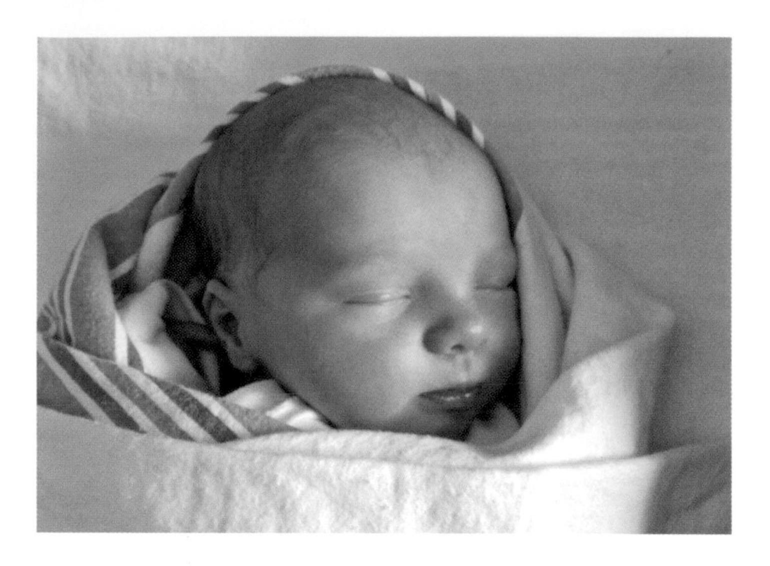

Husbands Joseph Vitale and Rob Talmas raced from their home in New York City to a hospital in Ohio in April 2013 to bring home their newly adopted son, Cooper (top). Months later, they discovered that the State of Ohio would only list one father's name on Cooper's birth certificate, deeming the other a legal stranger. Rob Talmas (bottom left) and Joseph Vitale lift their son, Cooper, in front of the U.S. Supreme Court in April 2015, the day their lawyers argued for marriage equality in the name of more than thirty plaintiffs from four states, including Cooper. (ABOVE Courtesy of the Talmas-Vitale Family, BELOW Courtesy of the Talmas-Vitale family, photo taken by Scott Fausett)

Pam Yorksmith (left) and her wife, Nicole (right), became Al Gerhardstein's plaintiffs when they learned that the State of Ohio would only list Nicole's name on the birth certificates for their two boys, Orion (left) and Grayden. "I do the four A.M. feedings," Pam Yorksmith told her wife, "but I am always less than." (AP Images / Gary Landers)

Grayden Yorksmith, four, sits on the steps of the U.S. Supreme Court the day before oral arguments in April 2015. His mothers had been married in California in 2008. In the hallway of their home, they kept a wall clock permanently set to 6:30 P.M. to commemorate the moment they exchanged wedding vows. (Courtesy of the Yorksmith family, photo by Vanessa Melendez)

Tennessee plaintiffs Sophy Jesty (left) and Val Tanco (right) stand with their lawyer, Regina Lambert (center), in front of the U.S. Supreme Court the day before oral arguments. Jesty and Tanco, married in New York, wanted to secure an accurate birth certificate for their daughter but the State of Tennessee would not recognize their marriage. (Courtesy of Regina Lambert)

Knoxville attorney Regina Lambert took her father, John Paul Lambert, to Cincinnati for a rally the day before the cases from Ohio, Tennessee, Kentucky, and Michigan were argued before the Sixth Circuit Court of Appeals. "It was initially difficult to accept that he had a gay child," Lambert, who is gay, said later. "Fortunately, we worked things out early in my life and his love for me allowed his heart to change." (Courtesy of Regina Lambert)

Kentucky plaintiffs Greg Bourke (second from left) and Michael De Leon (second from right) married in Niagara Falls in 2004 in front of their adopted children, Isaiah (far left) and Bella (far right). They led busy lives filled with church, soccer practices, and Boy Scouts meetings, but got involved in the case so that both fathers could be named on their children's birth certificates. In 2014 the family took a spring break trip to New York, where they toured the 9/11 Tribute Center. (Courtesy of the De Leon and Bourke family)

In August 2014, Greg Bourke, Michael De Leon and their children waited for oral arguments to begin at the Sixth Circuit Court of Appeals in Cincinnati. Plaintiffs from Kentucky, Ohio, Tennessee, and Michigan met for the first time and took turns passing around cell phones so that each family could get pictures on the historic day. (Courtesy of the De Leon and Bourke family)

Sixth Circuit Court of Appeals judge Martha Craig Daughtrey, one of three judges tasked with deciding the same-sex marriage cases, believed in the right to marriage equality and tried to convince the other two judges on the panel to vote with her. "These two men were in love," she emphasized, speaking to her colleagues about Jim Obergefell and John Arthur. (John Partipilo / The Tennessean. From The Tennessean, 2015-03-26 © 2015 Gannett-Community Publishing. All rights reserved.)

Early rulings by federal Judge Timothy S. Black required the State of Ohio to recognize the marriages of gay couples. "The marriage ban," he wrote, "embodies an unequivocal, purposeful and explicitly discriminatory classification, singling out same-sex couples alone, for disrespect of their out-of-state marriages and denial of their fundamental liberties." (Courtesy of the Cincinnati Bar Association)

Sixth Circuit Court of Appeals Judge Jeffrey Sutton, a leading conservative voice on the court, wrote the majority opinion ruling against the marriage plaintiffs from Ohio, Michigan, Kentucky, and Tennessee. "Not one of the plaintiff's theories . . . makes the case for constitutionalizing the definition of marriage," he wrote. This opinion would create a "split" among the nation's circuit courts and push the case toward a final ruling by the U.S. Supreme court. (AP Images / Evan Vucci)

Just after the historic decision at the U.S. Supreme Court on June 26, 2015, Al Gerhardstein and Jim Obergefell linked arms outside the building, where hundreds of cheering supporters waited to greet them. (Doug Mills / The New York Times/Redux Pictures)

Veteran attorneys Mary Bonauto (center) and Doug Hallward-Driemeier (right) were the two chosen to argue before the Supreme Court in April 2015; here Bonauto addresses journalists after their appearance. (AP Images / Cliff Owen)

Jim Obergefell greets another noted marriage equality plaintiff, Edie Windsor, at an American Civil Liberties Union reception in New York City a month after the historic *Obergefell v. Hodges* ruling. Windsor's case in 2013 forced the federal government to recognize state-sanctioned marriages among gay couples and provide all the federal benefits that come with marriage. It was on the day of the Windsor decision that Jim proposed to John Arthur. (Courtesy of Donna F. Aceto)

On the eve of oral arguments at the U.S. Supreme Court in April 2015, Isaiah De Leon listened to protestors make antigay declarations over a bullhorn. Instead of confronting them, he borrowed a rainbow flag and walked proudly past the protesters as his fathers watched. (Courtesy of the DeLeon and Bourke family)

Jim Obergefell speaks in Philadelphia after the Supreme Court decision in 2015. With same-sex marriage a constitutional right in all fifty states, Obergefell turned to new frontiers in the gay rights movement, including the need for a federal anti-discrimination law. (AP Images / Matt Rourke)

16

"ALL OF A SUDDEN THIS FAMILY WAS BORN"

THE PARENTS sat in the jury box, tense and silent. In the upper row, Joe Vitale clutched his husband's hand and tried not to worry about Cooper, home with an aunt in New York City. In the front row, Pam Yorksmith eyed three-year-old Grayden and the iPad wobbling in his lap, entirely unsure whether episodes of *Mickey Mouse Clubhouse* would get him through the next two hours.

"It's about families, Judge," Al said when the hearing in federal court was called to order just after ten A.M. on a brisk Friday in April. Outside, happy shoppers sampled pulled pork and gourmet popcorn at a street market on Fountain Square. Inside the courtroom, with its heavy drapery and wood walls, three teams of lawyers waited in leather chairs directly in front of Judge Timothy Black. Jim Obergefell sat alone behind them.

Judge Black eyed the parents and three pregnant women in the jury box to his left. "Well, let us proceed," he said.

"May it please the court," Al began. "This case is about families and the duty to treat families with same-sex married parents equal to families that have opposite-sex married parents. At this point, seventeen states plus Washington, D.C., permit same-sex marriages. Last December, you ordered Ohio to recognize same-sex marriages from those states in the context of death certificates when one of the [spouses] dies. In this case, we're asking you to order Ohio to recognize same-sex marriages from those states in the context of birth certificates—and in other contexts—"

"Excuse me," the judge interrupted. "In the prior case, the court ruled that Ohio had to recognize death certificates as it applies to Mr. Obergefell and the additional plaintiff. It was a limited ruling. Today, you seek a ruling striking down the constitutional amendment and the statute in whole as applied to everyone. Is that right?"

"That's correct with respect to marriage recognition."

Joe Vitale had read Judge Black's earlier ruling on death certificates, but now Al had elevated the dispute and was attacking the law itself. *A facial challenge.* The term had been new to Vitale, but it was clear that Al intended to ask the judge to take an unusual step, quashing a good part of a law that had been passed by Ohio's voters.

Al pointed to Document 66 in the case file, already more than a thousand pages thick with legal briefs and expert reports. It was Judge Black's first ruling on death certificates.

"Consistent with your finding," Al told the judge, " . . . the marriage recognition bans deny due process and equal protection, and the benefits of the existing marriages of people married in other states who are denied equal protection of the law when they come to Ohio. So you've already made a finding, although not a formal declaration or injunction, and that's where we're headed, hopefully, today."

"The Court recalls its prior decision."

" . . . What I would like to do today, Judge, is talk to you about whose rights are being denied." Al looked at the plaintiffs in the jury box. "They're all married in states where same-sex marriage is legal. All are establishing their families, bringing children into their lives."

Al thought of his three grown children and quipped, "Sometimes I wonder why they do that, but that's not for us to decide."

Judge Black, with two grown daughters, smiled, and said, "Well, the Court's prepared to rule on that issue because it's the greatest source of joy in the world."

Vitale told himself: Breathe. Breathe. He thought of Cooper again and how he turned mundane things into full color, the pigeon on the park bench, the sprinkle cookie in the bagel shop, every day new and animated.

Al went on smoothly. "So, if they were opposite-sex married couples, we wouldn't be here."

Al introduced Pam and Nicole Yorksmith and the two other female couples, pregnant through artificial insemination. "With married, opposite-sex couples, when they [give] birth due to artificial insemination, the spouse is presumptively on the birth certificate as the second parent."

He introduced Vitale, Talmas, and their adoption decree from New York, which stated that Cooper "shall be treated in all respects as the lawful child of the adoptive parents."

"So if they were an opposite-sex couple, we wouldn't be here because both of their names would be on the adopted child's birth certificate. But because they are a same-sex couple, Ohio will not honor their New York decree. Rather, Ohio requires them to choose between themselves and list only one as parent of the adopted child. Even though they are both parents, Ohio says, 'Only one of you can be on the birth certificate.'"

"My spouse wouldn't put up with that," Judge Black said.

"Well, they aren't either, and that's why we're here," Al replied. " . . . This is a serious problem. I mean, first, at a very basic level, at the basic level of human dignity, that dignity is denied by the way Ohio treats these couples. The first document that's welcoming their child into our civil society, the birth certificate, is going to be wrong. It's going to misidentify the child's parent only because the child has same-sex parents."

In the jury box, Vitale felt himself beginning to relax. Doubt had given way to a faint flicker of hope. "This is bizarre," Al went on. "In an era when we look for parents to step up and do their jobs, Ohio is saying, 'Well, we're only going to recognize one parent for children of same-sex couples.' That's outrageous and totally discriminatory."

Caught up in his argument, Al stopped reading from his notes, seventeen pages separated by topic in a three-ring binder. " . . . The state says we should honor the democratic process, which this court correctly disposed of in *Obergefell*. You cannot by popular vote deny constitutional rights to the minority. And the fact that it was voted on by the people doesn't make an unconstitutional law survivable."

Judge Black cut in. "As to the vote, the citizenry has trouble understanding that. Would you spend a little time on that and give me some analogies?"

At Beloit College in the late 1960s, Al had studied commentaries on the Constitution with Professor Harry Davis, who would spend forty-two years teaching political science. Al thought of an essay he had read in the professor's class, written by Alexander Hamilton when the founding fathers were promoting the ratification of the Constitution.

Federalist Paper 78, on the power of judicial review, was published in 1788.

> Independence of the judges may be an essential safeguard against the effects of occasional ill humors in the society. These sometimes extend no farther than to the injury of the private rights of particular classes of citizens by unjust and partial laws.

In the Constitution, Al learned, the oppressed held a trump card.

"We shouldn't have the tyranny of the majority," he said, his voice rising. " . . . And why is that? Because people can be whipped up into passion and then use their vote to deny to some insular minority, to some unfavored group, some flavor of the day that people are willing to discriminate against, whether it's blacks, whether it's illegitimate children, whether it's people from other countries, and they can vote away that group's rights. And from the very beginning of our country, the founding fathers, and mothers, said, well, we have to create a system where we can stop that, where we identify certain principles and those principles have to be honored even in the face of legislation passed by the majority."

" . . . So our Constitution and Bill of Rights provides a limited number of fundamental rights?" the judge asked.

"Right."

"And that's because we protect the minority from the oppression of the majority?"

"Right."

"And if one of the fundamental rights of the minority is violated in our country, the courts are required to stop it?"

"That's absolutely true," Al said. "And it goes to the central purpose of our civil society. We are a welcoming nation. We are a nation that encourages open thought, open speech, and encourages open communities. . . . Unless we accept people who are not in the majority, then we are denying the very core principles of our nation."

Al looked down at his notes again and delivered a parting line. "We do ask that... you declare the marriage bans unconstitutional in all applications in Ohio."

The state's attorney, Ryan Richardson, was nine months pregnant, and from the jury box, Joe Vitale's very first thought was, *You're going to stand up there and argue against parenthood?*

But Richardson, an expert in constitutional law, would focus on the state of the law, just like Bridget Coontz had done during the death certificate hearings. Since the beginning, Ohio attorney general Mike DeWine had made a strategic decision to focus only on the democratic process—respecting the will of Ohio's voters and their right to pass laws—steering clear of emotionally charged questions about the rights of gay couples and their families. It was a decision that Bridget Coontz, Richardson's supervisor, could live with, given that she personally supported same-sex marriage and had gay friends with adopted children.

When Coontz walked into the federal courthouse with Richardson that morning, she told herself, once again: *My clients got sued. My job is to defend my clients. The decision whether to defend wasn't mine—it was based on the state of the law.*

Coontz spotted Grayden Yorksmith in the jury box and briefly wondered whether her own four-year-old daughter could sit through a two-hour court hearing. Then she put on her game face, knowing full well that much of the room would accuse the state's lawyers of being antigay and antichildren.

"From the outset of this litigation, including throughout the complaint, plaintiffs limited this case to the context of birth certificates and sought relief for the specific plaintiffs involved in this lawsuit," Richardson said. "As plaintiffs acknowledge today, however . . . they ask this court for a sweeping relief invalidating the marriage recognition laws in all of their applications. This court should decline to do so. . . . Neither defendant nor the court has the opportunity to fully evaluate and analyze the potential impact of the sweeping relief that plaintiffs request."

Judge Black frowned. "Are you asking for a continuance?"

"We're not asking for a continuance, Your Honor. . . . We would note that, as this court is aware, plaintiffs sought an expedited case schedule here because of the alleged exigencies that related to the specific plaintiffs."

"That's lawyer talk for the fact that they're going to have children any day now?"

"With which I sympathize," Richardson answered.

"I wasn't going to comment," Judge Black said, "but you're going to get through this argument, is that right?"

"I think so. I think we're safe."

"Congratulations. You apparently are due in three weeks, correct?"

"That is correct."

"We checked that out before we permitted you to argue," the judge chuckled. "You can gather yourself. I'm sorry."

Richardson wanted the judge to see how broad the case had become and limit his ruling only to the plaintiffs in the jury box. "And as a result of those exigencies and the arguments that . . . plaintiff's counsel made about the narrow scope of this case," she said, "the parties agreed to, and this court approved, an extremely expedited case schedule that involved resolving this litigation entirely in less than two months."

"Every day there's a violation of constitutional rights gives rise to irreparable injury, is that true?" Judge Black asked.

"We respectfully suggest that they have not demonstrated that there is a constitutional violation," Richardson replied. " . . . Deference is particularly important here where we are dealing with a statute that has been enacted by the voters."

"I'm worried," the judge said minutes later, "that lawyers who argue that the will of the voters trumps all misrepresent the law."

"And, Your Honor, we don't mean to suggest that the will of the voters trumps everything in all cases, but we certainly believe that it is under what the courts have clearly said, an important consideration," Richardson replied. "Saying that the will of Ohio voters cannot be considered sort of assumes that the statute is unconstitutional, which we suggest it isn't. . . . The traditional definition of marriage is a conceivable rational basis that we believe that the voters were allowed to bear in mind at the time that they voted, and it does support the challenged marriage recognition laws here."

"Well, then how did the ban against whites and blacks marrying get struck down?"

"Again, Your Honor, clearly the law does not allow the will of voters to trump in all cases, and where a statute is determined to be unconstitutional, the fact that it was a popular enactment will not save an otherwise unconstitutional statute, as was the case there."

Judge Black wasn't satisfied. "Actually, I'm focused on the traditional definition of marriage. The traditional definition of marriage was that it wasn't between blacks and whites and then the Supreme Court struck that down."

"That's correct," Richardson replied. " . . . But in this case, with respect to the issue of same-sex marriage, the courts have repeatedly acknowledged that tradition is something that can be taken into account."

From the jury box, Joe Vitale cringed. He had come to despise the term *traditional marriage* because he knew that his marriage didn't qualify. He reached for his husband's hand again, thinking about their seventeen years together and the way he felt about his family on weekend mornings when Cooper climbed into bed between them, smelling like baby shampoo.

"Well, we've had full-blown oral argument," Judge Black told the lawyers. "I'm not going to render a decision from the seat of my pants. I am going to take it under consideration and will issue a significant, thorough written decision."

Vitale held his breath, unsure what would happen next. He looked at Al, who was looking at the judge. The room was still.

Seconds later, Judge Black said, ". . . I intend to issue a declaration that the Ohio recognition bans that have been relied upon to deny legal recognition to the marriages of same-sex couples validly entered in other states where legal violate the rights secured by the Fourteenth Amendment to the United States Constitution."

He went on, " . . . I anticipate issuing a permanent injunction prohibiting the defendants and their officers and agents from enforcing the marriage recognition bans in Ohio."

The judge promised a written decision within ten days. "Court prepares to recess," he said and stood up. "Godspeed."

Vitale could see his husband tearing up, and he looked away because he didn't want to cry. In the hallway outside the courtroom, Vitale found Al and pulled him aside. "Can Judge Black change his mind?"

"He can do whatever he wants," Al said carefully. "We'll just have to wait."

Twenty-eight years before the hearing, when Tim Black was still litigating civil disputes as a cub lawyer at a large Cincinnati law firm, he and his wife adopted their first child, a three-day-old baby they named Abby. Their second daughter, Emily, was born biologically four years later. Every year on April 14, the day that Abby Black's adoption was finalized, the family of four sang a John McCutcheon folk song called "Happy Adoption Day."

For out of a world so tattered and torn,
You came to our house on that wonderful morn
And all of a sudden this family was born

Looking at the pensive plaintiffs in the jury box, the two men clutching hands, the pregnant women and their wives, the toddler fixed on his iPad, Judge Black once again decided that fundamental constitutional rights had been violated by the whims of the Ohio electorate. He found the state's case devoid of any rational justification to support the ban on marriage recognition and the policy on birth certificates arbitrary discrimination based on sexual orientation.

Over ten days, he drew up a forty-five-page order ruling that Ohio's ban on marriage recognition violated the right to equal protection and due process. The Fourteenth Amendment, among the most litigated sections of the U.S. Constitution, was ratified in 1868 to address the rights of former slaves after the Civil War. Now it would address the rights of the gay community of Ohio.

On the first page of his order, Judge Black wrote in bold, underlined letters: "Ohio's marriage recognition bans are facially unconstitutional and unenforceable under any circumstances."

On the second page, he cited an opinion by the U.S. Supreme Court, rendered more than seventy years earlier, that protected students who chose not to recite the Pledge of Allegiance in school. "One's right to life, liberty and property, to free speech, to a free press, freedom to worship and assembly and other fundamental rights may not be submitted to vote; they depend on the outcome of no elections," the court wrote.

The marriage ban in Ohio, Judge Black found, "embodies an unequivocal, purposeful and explicitly discriminatory classification, singling out same-sex couples alone, for disrespect of their out-of-state marriages and denial of their fundamental liberties."

On April 14, his daughter's adoption day, Judge Black released his final ruling. On the last page, he included the words to the John McCutcheon song, which he still texted every year to Abby, who was married and living in California.

There are those who think families happen by chance . . .
But we had a voice and we had a choice
We were working and waiting for you.

Though Judge Black had announced during the hearing that he would side with the plaintiffs, his ruling officially overturning Ohio's marriage recognition ban topped the nightly news. During a press conference on the steps of city hall, surrounded by the mayor and half of the city council, Chris Seelbach declared, "Most Ohioans believe that gays and lesbians should be treated equally under the law."

Al wasn't surprised when the state announced it would appeal the ruling to the Sixth Circuit, where an appeal was already pending for the death certificate case. At the federal courthouse, Cincinnati's WLWT News Five filmed a handful of protesters gathered with signs that read SODOM BURNED and ETERNAL FIRE.

Inside his suite of offices on the eighth floor, Judge Black sat quietly at his desk, under a black-and-white photo of his family taken during his first campaign for municipal court judge in 1993. King Solomon had once said, "If you see oppression of the poor, and justice and righteousness trampled in a country, do not be astounded."

The courts were places of recourse. The judge had done what needed to be done.

PART FOUR

LEGACY

"Change will not come if we wait for some other person, or if we wait for some other time. We are the ones we've been waiting for. We are the change that we seek."

—Then Senator Barack Obama, 2008

17

THE REHEARSAL

AL GERHARDSTEIN, sweaty and defiant, jabbed the air with a ballpoint pen, looking at his notes and then back again at the semicircle of lawyers who were attacking his case against the State of Ohio. He had started off with what seemed like a simple argument, one that had already won a series of rulings in federal court. But now the questions were flying, and on this scorching July morning in 2014, Al paced at the podium, flipping the pages of his three-ring binder.

He had traveled to Nashville, Tennessee, for a moot court, a practice used by lawyers to simulate court proceedings and oral arguments, much like a dress rehearsal. Here, in a conference room of a local law firm, Al would practice his delivery before a team of supporters that would play the role of judges, exposing and exploiting weaknesses in his case. It would be a grueling, daylong strategy session, hashing out legal principles over turkey sandwiches and iced tea.

"We have four babies and Ohio is refusing to recognize the marriages of their parents," Al told the group. "That's true also of the *Obergefell* plaintiffs. All they want is for the death certificate to say 'married,' under marital status—"

"Look," one of the lawyers interrupted. "You've got two sets of very compelling plaintiffs. Their stories are heartrending, no question about it. But I don't understand. The federal government forces the states that don't want to do this? I don't think that your compelling clients make any difference."

Al put his hands behind his back and shot back, "Ohio has identified an entire group of citizens, married gays and lesbians, and walled them off. It says that their marriages don't count, that their families are unworthy and that their children are unequal."

The lawyers kept pressing.

"So if another state were to decide that they are going to allow polygamist marriages, does Ohio have to, as a constitutional matter, recognize their marriage?"

"How about this? Another state allows siblings to marry, and that couple comes to Ohio. Must Ohio recognize their marriage?"

"Are you saying that there is a fundamental right to the recognition of your marriage?"

Al smiled slightly at the skeptical attorneys with their legal pads and laptops. He knew the questions would only get tougher when he stood in two weeks before a three-judge panel at the Sixth Circuit Court of Appeals, which had jurisdiction over cases from lower federal courts in Ohio, Michigan, Kentucky, and Tennessee.

More than twenty attorneys had come to Nashville to prep because the Sixth Circuit had pending same-sex marriage appeals from all four states within its jurisdiction and had decided that attorneys would present their arguments back-to-back on the same day in its Cincinnati courtroom.

The governor of Kentucky was appealing two rulings, both issued after lawyers Shannon Fauver, Dawn Elliott, and their legal team filed suit. One ruling had struck down the state's ban on the recognition of same-sex marriage, just as Judge Timothy Black did in Ohio. But the second ruling was more sweeping: it ended

Kentucky's ban on same-sex marriage entirely, effectively requiring the state to issue marriage licenses to same-sex couples. Federal judge John G. Heyburn II had called the state's arguments in support of the bans "not those of serious people."

Tennessee was appealing a ruling by federal judge Aleta Arthur Trauger, who had written on behalf of the plaintiffs, "At this point, all signs indicate that, in the eyes of the United States Constitution, the plaintiffs' marriages will be placed on an equal footing with those of heterosexual couples and that proscriptions against same-sex marriage will soon become a footnote in the annals of American history."

University of Tennessee adjunct law professor Regina Lambert had helped launch the case, along with Abby Rubenfeld, a noted gay rights advocate and civil rights attorney in Nashville who had successfully toppled Tennessee's sodomy law in 1996. A partner in a prominent Tennessee law firm had also offered to help, but he pulled out when his firm decided it didn't want to be involved in a same-sex marriage lawsuit.

At the University of Tennessee, a second-year law student spoke up and told Lambert, "My dad will do it." Bill Harbison was a veteran civil litigator and third-generation Tennessee attorney.

"It's the right thing," Harbison said when his son called to tell him about the case. He enlisted three other attorneys from his firm to help. The team brought on a seventh attorney from Memphis, Maureen Holland, as well as the National Center for Lesbian Rights. They had three sets of plaintiffs, including Valeria Tanco and Sophy Jesty, veterinarians who had met at Cornell University, married in New York, and moved to Tennessee to become professors; and Sergeant First Class Ijpe DeKoe and Thomas Kostura, who had married in New York just before DeKoe began a yearlong tour of duty in Afghanistan.

Harbison knew Tennessee attorney general Robert Cooper Jr.

because their fathers had served on the state supreme court in the 1970s, so he decided to deliver a copy of the lawsuit personally, telling the attorney general, "I've got something for you. I thought I would come tell you myself."

In Michigan, attorneys Carole Stanyar and Dana Nessel had sued in 2011 on behalf of two nurses, partners for nearly a decade and the parents of three adopted children, two with special needs. Because of the state's marriage ban, the two women couldn't jointly adopt their children. Their legal team, which included criminal defense attorney Kenneth Mogill and Wayne State University Law professor Robert Sedler, took the case to trial, bringing in experts on subjects ranging from psychology to the history of marriage and discrimination. "Many Michigan residents have religious convictions whose principles govern the conduct of their daily lives and inform their own viewpoints about marriage. Nonetheless, these views cannot strip other citizens of the guarantees of equal protection under the law," federal judge Bernard A. Friedman had ruled.

Ohio and Tennessee had pushed only for the recognition of marriages performed in other states. Michigan was fighting for the freedom to marry itself. Kentucky had two cases, one pushing for marriage recognition and the other for the right to marry. Now all of the appeals were headed to the Sixth Circuit, the very path that had initially worried the ACLU and other national gay rights groups early on.

Just before the moot court, Al had learned the names of the three judges who would decide the case, and it was a majority-conservative panel. One of the judges was Jeffrey Sutton, a respected legal scholar who had clerked for Supreme Court justice Antonin Scalia, long considered an intellectual anchor for the high court's right wing.

Al was battling a rare case of nerves. A *Washington Post*/ABC News poll had found that a record-high 59 percent of Americans supported same-sex marriage. But Al worried about a decisive defeat in the Sixth Circuit at a time when marriage cases from other states were heading to more progressive circuit courts. Only weeks earlier, a panel of judges on the Tenth Circuit Court of Appeals in Denver had ruled in favor of same-sex marriage, the first appeals court in the country to find that voter-approved marriage bans were unconstitutional.

"Sometimes justice is really clear," Al told his son, Adam, who had joined Al's practice as an associate after moving home from Minnesota, where he helped defeat a ballot measure banning same-sex marriage. "The lines are defined and we all know what should happen. And then you lose. It can happen. It's happened to me."

Standing at the podium during the moot court, Al looked at the attorneys from Kentucky and Tennessee, who had decided to team up with Ohio to practice for the Sixth Circuit. Michigan was preparing separately. Susan Sommer, the Lambda Legal attorney who was working with Al on the birth certificate case, James Esseks, the ACLU attorney involved in the death certificate case, and Sam Marcosson, a law professor at the University of Louisville, would play the role of judges, questioning Al during his argument.

"Do you have a fundamental right to have your marriage recognized from elsewhere even if you do not have a fundamental right to marry within the state?" one of the lawyers asked.

Al knew it was the most important question of the day. Before the Sixth Circuit, he would have to argue for marriage recognition without undercutting arguments about the right to marry, which was the ultimate goal.

Al paused and rubbed his hands together. "If we have a right to marry in another state and that right is fully supported and legal, yes, there should be a right to carry it across state lines."

The lawyers were whispering around him. "Realistically," Susan Sommer said, "we would be allowing the floodgates to open in

Ohio if we were to say, 'Yes. Out-of-state marriages for same-sex couples get recognized.' "

Al responded quickly, "The issue is Ohio can't just decide today to target a group that is disfavored, that people dump on . . . and say, 'Oh, we aren't going to let those people have marriage.' That kind of use of state power is abusive. It's irrational. It is not a legitimate purpose."

Al's voice was rising. "There are four children that deserve to have two parents on their birth certificates."

While Al was in Nashville at the moot court, Isaiah De Leon was savoring the blissful final days of summer break in a shamrock-green bedroom cluttered with video games and movie posters. He knew that his family would soon drive from Louisville to Cincinnati to hear arguments in a court called the Sixth Circuit and that his name might be read out loud because his fathers were plaintiffs in one of the same-sex marriage cases from Kentucky. But Isaiah was thinking more about the start of his junior year in Catholic high school than the case that bore his family's name, which is exactly the way his fathers wanted it.

Michael De Leon and Greg Bourke fell in love thirty years earlier when it was still against the law to be gay in Kentucky. De Leon, with black hair and mocha-colored skin inherited from his Mexican ancestors, was raised on a farm in rural Kentucky after his father fought in Korea and Vietnam and then retired from the army. De Leon had been working on his bachelor's degree in agriculture at the University of Kentucky when he met Bourke, one year older and four inches taller, who had a large Irish and German family in his hometown of Louisville. In 1987, they bought a red brick Cape Cod just down the street from Our Lady of Lourdes Catholic church, where a priest baptized their daughter Bella, adopted at birth, and then Isaiah, who joined the family as a toddler a year later.

Isaiah never actually talked with his fathers about why his family was different, mostly because his life seemed so ordinary.

He ate dinner on a dining room table that his great-grandmother bought in the 1900s. He gathered corn and potatoes in a vegetable garden that had crept across much of their backyard. He played soccer on a team coached by one of his fathers and earned badges in a Boy Scout troop led by the other. Isaiah was heavily involved in their parish, and the only thing that came close to God in his devout family was Notre Dame football, which was celebrated regularly at tailgate parties.

Isaiah had always figured his fathers were married, but on Bella's fifth birthday in 2004, Bourke and De Leon took the children to Canada and legally married in a wedding chapel overlooking Niagara Falls. Bella, in a pale yellow dress with white roses, gawked at the waterfalls, lit at night in rainbow colors. Isaiah decided he had never seen anything so big.

Once, in the lunch line in elementary school, another student turned to Isaiah and said, "You don't have a mom. You have two dads and your dads are gay."

"What did you say?" Isaiah shot back. In the mostly white school, he had been teased for being biracial, but occasionally he had also been told that his fathers were "faggots." Shifting from foot to foot with his lunch tray, he was angry and embarrassed, but since he visited his birth mother several times a year, he said, "Well, I do have a mom and her name is Wendy and she drives a truck."

Isaiah was pleased with himself for not whopping the kid across the head and rushed home to tell De Leon. "It sounds like you handled that perfectly," his father said.

A more urgent problem came later, when the Boy Scouts learned that Bourke was gay and dismissed him as a troop leader. Isaiah was in ninth grade, and as the case drew headlines, he traveled with his family to Dallas to help deliver petitions to the headquarters of the Boy Scouts of America. Isaiah told reporters, "I've been scouting with my dad all these years. There's no reason for this prejudice."

Bella snickered at her big brother, standing in his olive green scouting uniform in front of the television cameras, but then she looked at her father Michael, shrugged, and said, "He did pretty good."

The experience showed Isaiah that there were more than 1.4 million people who understood and accepted his family because they had signed their names to a series of petitions condemning the Boy Scouts' policy on gay members and leaders.

At the end of Isaiah's freshman year in high school, his fathers called him into the living room with Bella for a family meeting. They had watched Disney movies and played Monopoly here, on the wood floor with an oriental rug that De Leon and Bourke had picked up in the 1980s. A crucifix hung near the leather sofa.

"Hey," De Leon said, looking at Isaiah and Bella. "We're making a federal lawsuit to recognize our marriage."

Isaiah turned to his sister, confused. "What do you mean?" he asked. "Your marriage isn't recognized?"

"You're already married," Bella added. "We went away and you got married."

"We did go away," De Leon said carefully. "The federal government has just now started to recognize our marriage, but our state does not."

De Leon and Bourke had never discussed the matter of birth certificates with their children because they didn't want to create unnecessary angst. Now, Bourke said gently, "We are both your parents, but only one of us is recognized as your legal parent. It creates a problem if something happens to one of us."

Isaiah and Bella nodded. "Would you consent to be plaintiffs?"

Bourke asked. "It will be public and everybody will know your family situation."

"Well, they already know," Bella said.

"We're with you," Isaiah said.

And so Bourke had called Shannon Fauver and Dawn Elliott, who would take the case pro bono. Like all civil rights lawyers, the lawyers would be paid by the defendants—in this case, the state of Kentucky—only if they won the case. And no one knew whether that would actually happen or how long it might take.

In August 2014, Isaiah and Bella drove with their fathers to Cincinnati to hear their case—*Bourke v. Beshear*—argued at the Sixth Circuit. "You cannot touch your phones," De Leon warned from the front seat of the car. "You have to sit there quietly and respectfully. You cannot nod off."

Isaiah knew there would be plaintiffs from other states at the courthouse, and as he settled in for the ninety-minute drive, he briefly wondered whether their stories would sound much like his own.

In a hushed Victorian house in the heart of downtown Knoxville, Sophy Jesty kissed her sleeping wife and slipped out of bed. It was just before 3:30 A.M., but it would take a good five hours to drive from Tennessee to the hearing at the Sixth Circuit Court of Appeals.

Jesty, a veterinarian with pale skin and cropped red hair, hated being away from her family even for one night, so she had decided to start the day before dawn and drive home that same afternoon. "I'll call you as soon as I can," she whispered to her wife, Val Tanco, who grunted good-bye. Jesty drove along a slick stretch of Interstate 75, quiet except for the truckers, and thought about the day ahead.

Until they moved to Tennessee, Jesty and Tanco had never had any plans to take up the issue of marriage equality. They met at Cornell University, where Jesty was finishing a fellowship in veterinary cardiology and Tanco, ten years younger, was completing a residency in animal reproductive medicine. It was a natural career path for Tanco, who had grown up in Argentina learning about cattle production.

They married in Brooklyn on a clear May afternoon in 2012 and had a ceremony in an old barn strung with white lights in the foothills of the Smoky Mountains. Tanco's father had been unsure how to respond ever since he learned that his daughter was a lesbian, but he flew in from Argentina for the ceremony. When he looked at Tanco in a white wedding dress, long brown hair hanging loosely around her shoulders, he asked Tanco's mother in Spanish, "Where should I stand so that I can walk Vale down the aisle?" He gripped his daughter's hand so tightly that his knuckles turned white.

Tanco and Jesty had worried about the move to the South, but they were drawn to the crowd in veterinary medicine at the University of Tennessee. Jesty applied first and when she was offered the job, she said, "The only way I can take this position is if you find a position for my partner. She's a reproductive specialist."

"Give me a few days and we'll see what we can do," the department chair replied, and then brought in Tanco through a spousal hire.

They bought a house near the university, and the day after the *Windsor* ruling in 2013, Tanco became pregnant through artificial insemination. They knew attorney Regina Lambert and her brood of poodles through the vet school, and after discussing the lawsuit, Tanco and Jesty signed on. They were about to become mothers, but without a court order, only Tanco would have legal rights to the baby. They sued in October, when Tanco was four months pregnant, and then waited, through doctor's visits and parenting classes, for a ruling from the judge. The baby was due in March, and Jesty wanted parenting rights in the hospital.

Finally, on March 14, the judge issued a preliminary injunction, ordering the state to recognize the marriages of Jesty and Tanco, and two other sets of plaintiffs. Lambert worried the state would try to "stay" the ruling during an appeal to the Sixth Circuit, so she started calling Jesty every few hours to check on the status of the pregnancy. "Any movement in the right direction?" Lambert asked, anxious to secure the birth certificate. "Do you think it's going to be today?"

Emilia was born two weeks later, eight pounds, six ounces, with black hair and olive-colored skin inherited from Tanco. "I'm one of your mommies," Jesty said, cradling the baby in her neck. The following day, they got a birth certificate listing both their names.

Four months passed in a blur of late-night feedings, and now Jesty was heading to Cincinnati, missing her wife and daughter and thinking about the three judges who would hear their case. Would they see a family making a life together in a rambling house in Knoxville or something else entirely, something strange and unworthy, banned by 1.4 million Tennessee voters?

18

HEARTS AND MINDS

THE ROWDY crowd wore red for love and spilled across the lawns of Cincinnati's Lytle Park, with its perennial gardens and towering statute of Abraham Lincoln, dedicated in 1917 by former President William Howard Taft. Jim Obergefell stepped lightly to a makeshift stage draped in a LOVE IS LOVE banner, hoping his voice wouldn't fail him.

On the eve of the hearing at the Sixth Circuit Court of Appeals, church leaders, city council members, and former Ohio governor Ted Strickland had come to the marriage equality rally in the city's historic district, where about seven hundred people swayed and prayed in the sultry August air. Jim had written a speech about John, and as dusk settled in Cincinnati, he looked out at the mass of supporters with rainbow flags. In a lineup of prominent speakers, he decided to start with "So who am I and why am I up here?"

And then he told his story. "Almost twenty-two years ago, I fell in love with John . . ."

The crowd quieted and Jim looked down at his notes, careful not to lose his place. "I'd watched this horrible disease rob John of every ability—to walk, to stand, to care for himself, to feed himself, to use his limbs, to speak more than a few words. Even with all ALS took from him, it couldn't take our love or his desire to do something for us, to help protect me once he was gone, to be legally married."

He went on. "The state says my relationship, my marriage, means nothing. That isn't right. It isn't fair. It's disrespectful. . . . I'm a husband, and now a widower. I'm not willing to give up my right to be either."

The applause was fierce and exuberant, and in that brief moment among strangers, Jim didn't feel quite so alone. But the next morning, he dressed alone for the hearing at the Sixth Circuit. He drove alone to Al Gerhardstein's law office. He sat alone in the courtroom, clutching a pink ticket that said PLAINTIFF and absently studying a carving of a bald eagle, wings spanned, that was perched over a semicircle of leather chairs reserved for the three judges.

Joe Vitale squeezed in next to him. Later, back home in New York, Vitale would tell his husband, "We all have someone except for Jim."

The room was packed with lawyers and plaintiffs from all four states. Isaiah and Bella sat between their fathers, who had carried a Kentucky state flag at the rally the night before. Greg Bourke, in his Boy Scouts uniform, had told the cheering crowd, "We have raised our Kentucky family together. We are Kentucky, and after thirty-two years together, we have earned the respect for our marriage."

Michigan plaintiffs April DeBoer and Jayne Rowse huddled with their attorneys. The nurses were raising three children but had been barred by state law from jointly adopting them. The courtroom was filled with babies, born to Pam and Nicole Yorksmith, Kelly McCracken and Kelly Noe, and Brittani Henry and LB Rogers in the months since Judge Black's ruling. DeBoer and Rowse walked over to see Orion Yorksmith, sleeping in a sling crisscrossed across Pam Yorksmith's chest.

Regina Lambert, one of the Tennessee lawyers, brought along her eighty-seven-year-old father, whose politics swung far right except in the case of same-sex marriage, perhaps because his daughter was gay. "It's like a vegetarian with a hamburger exception," Lambert teased. At the rally the night before, the retired General Motors insurance adjuster had put his hand over his heart and sung "The Star Spangled Banner" in a red plaid shirt with a WHY MARRIAGE MATTERS sticker on the breast pocket.

Al's cocounsel, Susan Sommer, had ducked into a drugstore before court to pick up a box of granola bars, which she handed out to plaintiffs and lawyers just before the three-hour hearing was called to order. Sommer sat down next to Al and was struck, as she always was, by the will of the families sitting behind her. Quiet dignity, she called it.

Though there were never certainties in court, by Al's count, the ruling would be split. Judge Deborah Cook, nominated to the bench by President George W. Bush, would likely side with the appealing states, while more liberal Judge Martha Craig Daughtrey would likely agree with Judge Black and find marriage bans unconstitutional.

But Jeffrey Sutton, the youngest of the three panelists and a leading conservative voice on the court, was far harder to gauge. The fifty-three-year-old jurist, a visiting lecturer at Harvard Law School, believed that limiting the role of the federal government in the affairs of states was a critical tenet of the Constitution. But he had also cast the deciding vote three years earlier to uphold the constitutionality of President Obama's Affordable Care Act. Al studied the thoughtful judge, who was looking out across the expanse of the courtroom, and wondered whether his vote would decide this case, too.

Michigan solicitor general Aaron Lindstrom stood up to present his argument first, walking across the courtroom's red carpeting to a wooden lectern directly in front of the three judges. He started in a predicable place, but Al knew the argument would appeal directly to Judge Sutton.

In Michigan in 2004, 2.7 million voters had passed a ballot initiative banning same-sex marriage.

"It is a fundamental premise of our democratic system," Lindstrom began, "that the people can be trusted to decide even divisive issues on decent and rational grounds, and that's what this case is about. It's about who gets to decide what the definition of marriage is, not what that definition must be."

A former army cavalry officer and graduate of the University of Chicago Law School, Lindstrom had a delivery that was matter-of-fact and practiced, even when Judge Sutton intervened.

"Is it difficult to adjust state laws on marriage, divorce, anything else, or is it really pretty simple, you just now include this new group?" Judge Sutton asked, his voice soft and level.

"It would have widespread impact," Lindstrom replied. "I'm not quite sure exactly how all of those would play out—"

"What would they be? That's the question," Judge Daughtrey said with a faint southern accent. She had gone to Vanderbilt Law School in the 1960s when women were becoming nurses and teachers and was only the fourth female judge to serve on the Sixth Circuit.

She asked again: "What would that be?"

"Well, as far as changing all of Michigan's law about marriage? In the big picture, one of the things that could happen if it were changed, there would be no institution in Michigan that would say it's important to have both a mother and a father. So in terms of societal impact, I think there might be harm, which is that there would be nothing to say it's important for fathers to be there and mothers to be there, and mothers and fathers bring different things to the table."

Al cringed, thinking about the parents and children he represented, sitting behind him in the roped-off section for spectators, and was momentarily grateful that the State of Ohio had avoided such divisive arguments in Judge Black's courtroom.

"Do you honestly think that's what's happened in the states where same-sex marriage is now valid?" Judge Daughtrey pressed.

"I think it's too early to tell, Your Honor. It's only been ten years since the first state passed it and—"

"But we're now to something beyond twenty-five percent of the jurisdictions in the country, and maybe more than that in terms of population as a whole, and it doesn't look like the sky has fallen in."

"I think the point is that it's too early to tell. You're changing such a fundamental bedrock of society in just ten years. That's not even a single generation of children, so I don't see how it could be possible to access the outcome on children."

Judge Sutton listened quietly to the exchange, his expression unreadable. But he quickly stepped in when Carole Stanyar, one of the attorneys for the plaintiffs of Michigan, came to the lectern.

"I know that there's many significant benefits, some of them monetary, that get extended to same-sex couples if you win here, and I think that's significant," the judge said. "But I have to believe, based on the briefs, that the most important thing is respect and dignity and having the state recognize these marriages the same way heterosexual marriages are recognized, and if respect and dignity are critical, are the key elements here—"

Al could see where the question was going and tensed. He knew that Judge Sutton believed the courts were meant to enforce laws, not create them, and that the voters were ultimately in the best position to decide law and policy.

The judge sighed softly and continued, "Maybe it's just something I'm missing, but I would have thought the best way to get respect and dignity is through the democratic process.... Nothing happens as quickly as we might like, but I'm just curious how you react to that point."

It was the first significant challenge of the hearing, and Stanyar, a veteran criminal defense lawyer who had sold her house to fund the case even after some gay rights advocates suggested it was unwinnable, responded quickly. "The Michigan marriage amendment gutted the democratic process in Michigan. Voters can no longer appeal to their legislators..."

The judge was unconvinced. "Aren't you optimistic that Michigan voters, if another initiative were put in front of them, it certainly would be a different vote and it might well be a different outcome even today?"

"You would have to come up with the signatures of ten percent of the total number of voters that, you know, were in the last general election. It's very cost prohibitive for a disfavored minority to be doing that."

In 2012, gay rights groups spent $15 million in Minnesota to defeat a constitutional amendment banning same-sex marriage, and another $15 million in the state of Washington. In Michigan and Ohio, with larger media markets, advocates had estimated the costs would likely double.

"The goal is to change hearts and minds...," Judge Sutton went on. "Isn't it worth the expense, and don't you think you are most likely to change hearts and minds, through the democratic process than you are with the decision by five justices of the U.S. Supreme Court?"

Stanyar fired back. "Fundamental constitutional rights may not be submitted to popular vote. They depend on the outcome of no election."

"... My question is assuming you can win on this," the judge said quietly. "I'm asking you a question: Why do you want this route? It's not one hundred percent obvious to me why it's the better route. It may be the better route for your clients, and as a lawyer you have to keep the focus on that, but it's not one hundred percent obvious to me that it's the better route for the gay rights community. That's not obvious to me."

Al shifted in his seat, anxious for his chance to appeal to the panel. Judge Sutton was a deliberate, astute speaker and writer, but in this case, Al believed the judge had slipped across a line, acting as some sort of campaign adviser rather than a jurist charged with deciding whether fundamental rights had been trampled by popular will.

Al decided he had his work cut out for him.

State Solicitor Eric Murphy would represent Ohio, and he, too, started by arguing for the democratic process. "The fundamental question in all of these cases is the same, and that fundamental question is not whether Ohio should recognize same-sex marriage but who should make that important decision of public policy on behalf of the state."

"What implementation problems arise if the plaintiffs win?" Judge Sutton asked.

Murphy had argued major appeals for the global law firm Jones Day before being tapped by Ohio attorney general Mike DeWine. He replied, "I don't know if it's an implementation problem, but I think that it would certainly require a legislative response. For instance, birth certificates have 'father' and 'mother.' I mean, maybe it should be changed to just 'Parent 1' and 'Parent 2.' I mean this is just a pragmatic response—"

"It's a pragmatic question," Judge Sutton said.

"I would imagine those things would happen throughout the Ohio Revised Code, where there's reference to husbands and wives. . . ."

"Other statutes about divorce, adultery, all that," Sutton said, "all I'm hearing you say is that, well, yeah, you would just have to use 'spouse' or—"

"'Spouse' or 'parent,'" Judge Daughtrey added.

"That's all that would have to change. Nothing else?" Judge Sutton said.

"... Frankly, Your Honor, I didn't anticipate the question, so I didn't review the revised code all that closely myself."

"You didn't bring it with you?" Judge Sutton said to a discreet round of laughter.

"Maybe I should have," Murphy replied. "I do think that there's no doubt that it would require new laws being passed—"

"Or forms being reprinted, perhaps," Judge Daughtrey said.

"It's just too new today," Murphy said minutes later. "The law is always about drawing lines, and at one point, maybe it becomes an irrational idea to proceed with caution. But I don't think we're there yet."

Murphy was wrapping up when Judge Daughtrey leaned forward. "Do you have any knowledge of how many years it was from the start of the campaign until the Nineteenth Amendment, when women achieved the right to vote? Are you familiar?"

"Ah, I'm not, Your Honor. I'm sorry," he said.

Al's law partner, Jennifer Branch, whispered, "Eighty years." Branch's grandmother had cast her first vote in the early 1930s, backing Franklin Roosevelt even after her husband, a Republican, refused to drive her to the polls.

Judge Daughtrey, who kept a first edition of the four-volume *History of Woman Suffrage* signed by Susan B. Anthony in her chambers, said sternly, "If I told you that it took seventy-eight years of crossing the desert back and forth, back and forth, trying to achieve it through the democratic process, would you be surprised?"

"Well, not with respect to the United States Constitution, because the United States Constitution sets a very high bar for constitutional amendments," Murphy replied.

"No, no, no, no, no," the judge said quickly. "I'm talking about going into every state in the country, every city, every school board election for seventy-eight years and trying to get enough going to convince the legislatures to adopt or extend the vote to women, seventy-eight years of it. And would you be surprised to find out it didn't work and it took an amendment to the Constitution to finally achieve that after seventy-eight years?"

"Yes, there's no question that the U.S. Constitution is very different—"

"No, no, no. That's not my question. You—"

"But with respect to the Ohio constitution—"

"Excuse me. You're not getting the point. The point is you want to do this democratically, state by state, legislature by legislature, municipal government by municipal government, as far as I know, and it just doesn't always work."

"Well, that's, that's—"

"It doesn't always work—seventy-eight years to get women just the right to go to the polls and vote." The judge's voice softened. "That's all. You don't have to respond. It's okay. I just thought you'd like to know that in case you are ever on *Jeopardy*."

Watching the exchange, Kentucky lawyer Shannon Fauver told herself, "She's on our side."

Fifteen-year-old Bella De Leon, sitting beside her brother and fathers, was so taken with the judge that she thought: I'm going to go to law school.

The hearing was nearly halfway over when Judge Sutton finally gestured to Al. "Okay. Mr. Gerhardstein."

Al straightened his shoulders, stood up, and passed Eric Murphy on the way to the lectern. "Ohio issues a birth certificate that names only one member of each couple as the parent and denies recognition as a parent to the other. That's a real serious harm," Al said. "Ohio also says to the surviving spouses in *Obergefell,* 'You must accept a death certificate for your loved one that's wrong, one that does not say you were married even though you are, and one that leaves blank the spot where your name should go as the surviving spouse.'"

Judge Sutton interrupted. "Can I just ask you a framing question, and I'm fearful it's a little simplistic, but I'd love to hear your reaction to it because we do have all these cases and we have all these issues."

Al waited.

"My rather simplistic way of looking at it is isn't the first question whether a state can decide, for its own purposes, its own citizens, whether to recognize same-sex marriage, and if it decides that it's not going to do that for now and if the U.S. Constitution . . . permits that choice, I guess it seems really odd to me that they can be told, 'Okay, even though you can make that choice for your own citizens, if someone comes from another state, that public policy choice doesn't bind you.'"

" . . . Once you're married," Al answered, "that attaches all kinds of vested rights. You have important parenting rights. You have important child-rearing rights that have been recognized by the Supreme Court. And for history, that's been transportable across state lines. This . . . involves people who have a history of discrimination and an issue that's very personal and carries with it very important rights and—"

"I agree that there's a history of discrimination. I don't think there's any doubt about that," Judge Sutton said. "I guess what is not so obvious to me is a history of discrimination when it comes to access to marriage. That seems to me a much more recent phenomenon and just a reflection of the current times and a new sensitivity on both sides of the debate."

"The deal that these couples made when they got married in New York, California, Massachusetts, Maryland, and Delaware," Al replied minutes later, "was that they would have a marriage that they could carry—"

Judge Cook, once a possible nominee for a seat on the U.S. Supreme Court, cut in. "Can we add into the logic of this that they were well aware that they were moving to a state where same-sex marriage was not recognized?"

"Your Honor," Al said. "We are in a situation where the democratic process has evolved, okay—"

"But I think it goes to the thinking that you proposed—"

"Right, but there's not like a contributory negligence defense to a constitutional right," Al said. "I mean, either your marriage is transportable or it's not. They got married because they were in love. They didn't get married trying to think of well, you know, 'Where can I go here?' and 'Where can I go there?' They do expect that their marriage will be transportable. That's a reasonable expectation. In fact, forty-four percent of the people in this country now live in a state where same-sex marriage is available, where the freedom to marry has been recognized, and that includes those twenty to twenty-one states where the deal is done, where there's no more appeals pending and so on."

"That could go both ways, wouldn't you agree?" Judge Sutton asked. " . . . On the one hand, it helps you in the sense that maybe you're getting to some tipping point where it's just outlier states and the courts step in. On the other hand, it suggests the democratic process is working and indeed working effectively and very quickly from your clients' perspective."

" . . . It's been a long process of development," Al replied, "but, you know, Judge, what I'm suggesting is that the ultimate role of the federal court is to keep states from denying the liberty to certain citizens, and here when you've got citizens who have a liberty interest—their marriage already exists, their marriage is done—and they've now got children and those children deserve to have two parents, and the state is now saying, 'Because of our commitment to democracy, we're just going to say no to you. And we're just going to wait for you to . . . reverse our constitutional amendment and, you know, we'll see you in a few years when you can pull off that kind of fundraising and that kind of democratic action.'

"The reality is that these rights are very, very profound," Al continued, "and we know from Supreme Court case law that a marriage is a very significant thing. It's solemn. It's precious. It's got all these attributes that allow you to have the relationship with your children and with your spouse. And this can't be just subject to vote."

"It shouldn't just be subject to vote," Judge Sutton said. "But I'm just curious why you're so sure about the better path. In other words, let's say the gay community gets to pick the path. You can get your Supreme Court decision in June of next year or you can have five years to change hearts and minds through democracy in the remaining twenty-nine states. It's just not obvious to me what's the best path."

"Well, I'm trying to suggest a constitutional path—"

"I get it. The assumption of the question was that you can have either one. That's the assumption of the question. It's just not obvious to me why the Supreme Court ruling by five justices in June of 2015 is the better path for the community—not necessarily your clients, the community at large. Changing hearts and minds happens through democracy much more effectively than happens through court decisions."

Al lowered his voice. "I understand, judge, but I represent four couples. Their kids deserve two parents. They deserve them today."

Al turned around, motioning to the families behind him. "You've got the non-birth mothers of these three babies saying, 'I am a parent. Sue me if my kid doesn't get my support. Call me if my kid doesn't show up for school. Prosecute me if there's neglect of my kid.' And Ohio is saying, 'No. We don't want that. We'll let this kid only have one parent but if you're an opposite-sex [marriage] kid, then you will have two parents.' That's a super harm to these children. And that's part of why the matter is urgent."

"I've been married to the same woman for forty-two years, three great kids," Al concluded. "The law is rigged in my favor."

Two hours had passed, and since the arguments in Kentucky and Tennessee were similar to those in Ohio and Michigan, the remaining attorneys had been given only fifteen minutes apiece to address the court. But Al was anxious to hear from the attorneys for the two states, which had suggested that bans on same-sex marriage were rational because heterosexual couples could procreate, promoting stable families.

Attorney Leigh Gross Latherow, a civil litigator who was hired by Kentucky's governor after the attorney general declined to defend the marriage ban, said, "Kentucky has said that perpetuation of the human race leads to stable birth rates which, in turn, leads to a strong economy."

" . . . Marriage doesn't mean you have to procreate," Judge Daughtrey interrupted. "There is a right not to procreate."

"There is a right not to procreate, Your Honor. The question is when you're looking at a governmental benefit, which is a marriage license . . . is the group who gets the benefit, and here, traditional man-woman couples get the benefit, does that benefit further the state interest? And it does further the state interest."

"Can you tell me how, though? We're back to the old 'What is it about same-sex marriage that will stop procreation in the state of Kentucky?' "

From his seat near the middle of the courtroom, Greg Bourke looked at Bella and Isaiah. Behind him, Joe Vitale thought of Cooper, home in New York. Both men were thinking the same thing: Why would their marriages stop heterosexual couples from having children?

"The state interest is in procreation," Latherow continued later, "and we believe that couples who are married procreate. It's really very basic—"

"Well, couples who aren't married, in fact, procreate, too, and more and more couples these days are not getting married and are procreating," Judge Daughtrey countered.

"They do," Latherow said. "There's no question about that."

"So how does the law advance procreation?"

"We believe that in the confines of marriage that procreation occurs, just as it does outside. It doesn't mean that it doesn't happen outside. The law doesn't have to be drawn with mathematical certainties."

"Okay. So I can put this down that what the Kentucky law does is cause procreation?"

"It furthers the interest in procreation. Yes, Your Honor."

"Okay, but wait, wait, wait, wait. I think we're getting circular now. How does it foster procreation, and you're telling me it does so because married people procreate?"

"That's correct."

"Is that what you said?"

"Same-sex couples—biology—same-sex couples cannot procreate. They can perhaps do artificial insemination. They can perhaps have a surrogate. But that's not a procreation of that couple . . ."

"Okay. I got it. Married, opposite-sex couples procreate?"

"Yes, Your Honor. . . . And that advances the legitimate interests of the commonwealth."

Attorney Bill Harbison would speak for the plaintiffs from Tennessee, including veterinarians Sophy Jesty and Val Tanco. "For the life of me," he told the judges, "I cannot see a logical connection between the effect of these laws, which is to exclude recognition of a category of marriages, and promoting anything having to do with procreation, one way or the other."

When the hearing ended, Al walked back to his law office, knowing it could be weeks or months before the panel delivered an opinion. Joe Vitale flew home to his husband and Cooper in New York City. Bella, Isaiah, and their fathers drove home to Louisville. Regina Lambert and her father drove home to Knoxville.

Susan Sommer couldn't get a flight back to New York City until the next morning so she walked to her hotel, enjoying the afternoon sun after a long morning in court. Once, at a children's birthday party, another mother had mentioned that she was taking her young son to counseling because she feared he might be gay and worried about a difficult life ahead of him. What we need to do, Sommer thought angrily, is not change the child but change the world around him.

Near the hotel, she passed boisterous groups of people heading downtown. Sommer smiled as they walked by. The crowd wore red, on this day for the love of Cincinnati baseball.

But it was love nonetheless.

19

BEING THERE, AGAIN

THE THREE judges gathered in the robing room right after court adjourned and shut the door behind them, scattering a battery of anxious law clerks to their upstairs offices. Martha Craig Daughtrey hung up the black robe she had put on before the hearing and sat down at the conference table with her notes, first on Michigan, then Ohio, Kentucky, and Tennessee.

Alone in the wood-paneled room, named for a closet of robes used by the judges on the Sixth Circuit, judges Daughtrey, Cook, and Sutton would discuss the case for the first time. Then they would take a secret vote. If there was dissent, it was considered polite to speak up so that the presiding judge, in this case Jeffrey Sutton, could quickly decide who would write the court's majority opinion, a massive research and writing assignment.

Judge Daughtrey knew that this would be her best chance to press the case for same-sex marriage because judges weren't supposed to lobby colleagues individually after the vote. She had come to the hearing hoping that Judge Sutton would take her side and find the marriage bans unconstitutional. But his comments about the democratic process had been telling, and when he looked across the conference table and said that he was thinking about siding with the states—reversing the lower-court rulings—Judge Daughtrey was deeply disappointed.

She would try to move him anyway. She recounted the case of Jim Obergefell and John Arthur. "Jeff," she said, pleading with the younger judge, "these two people were in love."

She walked back to her office alone a few minutes later, on the losing side of a 2–1 vote. Her disappointment had given way to indignation, and when she opened the door to her office and looked at her law clerks, she sighed and said, "I hoped he wouldn't want to go against history."

Her feelings would spill into a blistering dissenting opinion that she would author in the coming weeks, starting with a quote by Benjamin Cardozo, who had succeeded Justice Oliver Wendell Holmes Jr. on the U.S. Supreme Court in the early 1930s: "The great tides and currents which engulf the rest of men do not turn aside in their course and pass the judges by."

Fifty years earlier, in a first-year law class at Vanderbilt with 127 men and three women, Martha "Cissy" Daughtrey had discovered firsthand how slowly equality came to the weak and disenfranchised.

She had grown up in the 1950s amid the steel mills of Middletown, Ohio, and was schooled on a high school scholarship at the National Cathedral School in Washington, D.C. In 1961, she married a political reporter for Nashville's morning newspaper. When she showed up at law school in a green corduroy jumper, the only maternity outfit she could afford on an eighty-five-dollar-a-week journalist's salary, her faculty adviser never spoke to her again. A constitutional law professor asked her to stand up in class and recite a 1961 Supreme Court case, *Poe v. Ullman*, which had kept intact a Connecticut law that banned the use of contraception. It could have been a coincidence, but Daughtrey didn't think so.

There was no money for childcare, and there were no day-care centers in Nashville anyway, so she left law school when her daughter was born. She eventually found a preschool that admitted toddlers who were potty trained and spent the summer of 1966 hovering over a toilet. "Look," she told her daughter, "I don't want to put any pressure on you. But if we can get this done in the next couple of months, your mama can go back to law school."

In her third year at Vanderbilt she started hunting for a job in bustling Nashville, with its Country Music Hall of Fame and downtown auditorium that hosted the *Grand Ole Opry*. But there were only three practicing female lawyers in town, and though Daughtrey was in the top 5 percent of her class, no Nashville law firm would hire her. She applied to a local bank, thinking that trust work would be considered an appropriate job for a woman, but the vice president said the bank hired men only. She promptly withdrew the seventy-five dollars in her savings account and stomped out.

She got a job in the U.S. Attorney's Office, prosecuting criminals before all-male juries, and later returned to Vanderbilt to become the university's first female law professor. As the only woman in the faculty lounge, she often found herself debating male professors about the burgeoning women's movement. She was nominated to the Tennessee Court of Criminal Appeals, the first woman to serve on any court of record in the state. Once, a male lawyer in the throes of an oral argument had looked up at her on the bench and said, "Honey, I'm so glad you asked me that question."

"I'm not offended," Judge Daughtrey told him later, "but I just wanted to point out what you had done because if you ever call Judge Joe Duncan over here 'honey,' I think he might be offended."

In 1993, as her daughter prepared to graduate from Vanderbilt

Law School, Daughtrey was nominated by President Bill Clinton to a seat on one of the most powerful courts in the country, the U.S. Court of Appeals for the Sixth Circuit. The same-sex marriage cases coming out of Ohio, Michigan, Kentucky, and Tennessee felt to the judge like being there all over again, in a place and time when it was acceptable to marginalize an underclass because that was the way it had always been: us and them.

In her office after the vote in the robing room, she glanced at her books on the suffrage movement. In that low moment alone, knowing that the move toward marriage equality had just suffered a significant setback, she thought about the bumpy, long road to women's rights. She felt certain of the similarities. Seventy-eight years spent waiting on the democratic process to work, Judge Daughtrey decided, had only seemed like a long time to the women who were waiting.

On the first Thursday in November, attorney Regina Lambert climbed into her BMW at the University of Tennessee and turned on her cell phone. She had been waiting weeks for news of a decision by the Sixth Circuit, checking her texts, checking her e-mails, checking the Internet, waiting and waiting again. Other courts had been drawing headlines, and Lambert could barely process the news. In September, one month after the Sixth Circuit hearing in Cincinnati, the Seventh Circuit Court of Appeals struck down marriage bans in Indiana and Wisconsin—the third circuit to rule on the side of same-sex marriage.

In early October, the U.S. Supreme Court declined to hear appeals from any of the three circuits. Rulings in favor of same-sex marriage in Idaho, Nevada, and Alaska followed in startling succession. By November 6, when Lambert turned off her cell phone to teach a legal writing class at the University of Tennessee, same-sex marriage was legal in thirty-five states. She had married in one of them right after the *Windsor* ruling, traveling to Vermont to exchange vows with her partner of twenty-five years before returning home to Tennessee, one of the remaining fifteen states with marriage bans.

She was pulling out of the parking lot when her cell phone started buzzing. She looked down, surprised to find dozens of new e-mails. The very first one was the decision from the Sixth Circuit, so she pulled over to the side of the road and, with shaking hands, scrolled to the last page of the ruling.

She settled on two stark words: We reverse.

Lambert sucked in her breath. She stared at her phone, blinking rapidly and fighting nausea. It felt like someone had socked her right in the gut, the same feeling she'd had in the courtroom back in August when Judge Sutton had talked about changing hearts and minds through the democratic process, no matter how long that might take. Sitting next to her father, Lambert wanted to stand up and shout, "I really don't care about hearts and minds. These are life rights. It's *now*."

She knew she needed to call plaintiffs Val Tanco and Sophy Jesty to tell them that the Sixth Circuit had overturned the lower-court rulings and upheld the marriage bans, but she struggled to find the right words. Seven months earlier, Lambert had raced to the hospital to hold their newborn daughter, swaddled in a white fleece blanket with a grinning giraffe, and decided to stay all day to be sure that both mothers' names were listed on the birth certificate, in compliance with the order by U.S. District Judge Aleta Arthur Trauger. But the order had been preliminary, and now the Sixth Circuit had invalidated it.

Lambert started reading the forty-two-page opinion, written by Judge Sutton and seconded by Judge Cook, sealing the 2–1 vote.

> Not one of the plaintiff's theories . . . makes the case for constitutionalizing the definition of marriage and for removing the issue from the place it has been since the founding: in the hands of state voters.

Judge Sutton had cited a watershed 1971 lawsuit in Minnesota, filed on behalf of two gay men who were denied a marriage license. The Minnesota Supreme Court had ruled that limiting marriage did not violate the Constitution, and the U.S. Supreme Court, when asked to consider an appeal, had rejected the case with a single line: "for want of a substantial federal question."

The Supreme Court, Judge Sutton argued, had never undercut that precedent, not even in the *Windsor* decision, which ordered the federal government to recognize same-sex marriage but did not directly address whether state bans violated the Constitution.

Lambert skimmed the detailed sections in the middle of the ruling, where Judge Sutton had described the importance of judicial constraint and the sanctity of the democratic process.

> A dose of humility makes us hesitant to condemn as unconstitutionally irrational a view of marriage shared not long ago by every society in the world, shared by most, if not all, of our ancestors, and shared still today by a significant number of the states.
>
> . . . For all of the power that comes with the authority to interpret the United States Constitution, the federal courts have no long-lasting capacity to change what people think and believe about new social questions. . . . Isn't the goal to create a culture in which a majority of citizens dignify and respect the rights of minority groups through majoritarian laws rather than through decisions issued by a majority of

> Supreme Court justices? It is dangerous and demeaning to the citizenry to assume that we, and only we, can fairly understand the arguments for and against gay marriage.

He had used forty-one words to sum up his answer to the question posed by the plaintiffs of Ohio, Kentucky, and Tennessee: Does the Constitution prohibit a state from denying recognition to same-sex marriages conducted in other states?

> If it is constitutional for a state to define marriage as a relationship between a man and a woman, it is also constitutional for the state to stand by that definition with respect to couples married in other states or countries.

He closed with a final thought.

> When the courts do not let the people resolve new social issues like this one, they perpetuate the idea that the heroes in these change events are judges and lawyers. Better in this instance, we think, to allow change through the customary political processes, in which the people, gay and straight alike, become the heroes of their own stories by meeting each other not as adversaries in a court system but as fellow citizens seeking to resolve a new social issue in a fair-minded way.

Lambert dialed Jesty and Tanco. "We got the news," she said quietly.

That night, at home with their daughter, three cats, and two dogs, rescues named Biscuit and Carlos, Jesty felt as if Judge Sutton's words were directed at her, and worse, her family. For the first time, she felt completely disconnected from the state of Tennessee. Would the government try to take away Emi's birth certificate? Tanco and Jesty had decided they would never give their copy back. But now even Emi's last name—the same as Jesty's—could potentially be challenged since Jesty, in the eyes of their state, had no legal claim to the baby.

They sat on the couch after Emi fell asleep in a nursery painted turquoise. It would get colder soon and they would start picking up leaves from the oak and walnut trees in the backyard. They had fallen in love with the hundred-year-old Victorian house during one of their early visits to Tennessee in 2011, and they often sat outside late into the evening, drinking beer on a bright-yellow porch swing.

"We lost," Jesty said. Her face was streaked with tears.

Years earlier, when Tanco was struggling with a tough climb on a hike in the mountains of upstate New York, Jesty had traced the contours of her face, cheeks, eyes, nose, mouth. Jesty was just like that, steadying and serene, and over time she had brought Tanco a great sense of comfort.

"I'm sorry, love," Tanco said, and she meant it for all three of them.

The news about the Sixth Circuit made its way to New York, where Joe Vitale turned to his husband. "Don't worry. We're going to take this to the Supreme Court."

"What are you talking about?" Rob Talmas said, his voice grim. "We're nobodies."

The news made its way to Cincinnati, where Judge Tim Black sat alone in his office, reading a decision that had overturned his own. He wondered whether Judge Sutton had made a tactical move, creating a "split" among the country's circuit courts to compel the U.S. Supreme Court to take up the issue of same-sex marriage.

Judge Black knew that as the court of last resort, the Supreme

Court received about seven thousand requests every year to hear cases but would consider only a fraction of them, most often those of exceptional national importance or those that had created conflict and confusion in the lower courts. Just weeks earlier, Supreme Court justice Ruth Bader Ginsburg had told a group at the University of Minnesota Law School that there was no urgency to take up the issue of marriage because the circuit courts had all ruled the same way. But now the Sixth Circuit changed all that.

Regina Lambert was also considering Justice Ginsburg's prescient remarks. "We're the split," Lambert thought that first Thursday in November. Then, to make herself feel better, she read the dissenting opinion in the case, authored by Judge Daughtrey after the 2–1 vote.

> Instead of recognizing the plaintiffs as persons, suffering actual harm as a result of being denied the right to marry where they reside or the right to have their valid marriages recognized there, my colleagues view the plaintiffs as social activists who have somehow stumbled into federal court, inadvisably, when they should be out campaigning to win "the hearts and minds" of Michigan, Ohio, Kentucky and Tennessee voters to their cause.
>
> But these plaintiffs are not political zealots trying to push reform on their fellow citizens. They are committed same-sex couples, many of them heading up de facto families, who want to achieve equal status . . . with their married neighbors, friends and coworkers, to be accepted as contributing members of their social and religious communities and to be welcomed as fully legitimate parents at their children's schools. They seek to do this by virtue of exercising a civil right that most of us take for granted—the right to marry.

> ... More than 20 years ago, when I took my oath of office to serve as a judge on the United States Court of Appeals for the 6th Circuit, I solemnly swore to "administer justice without respect to persons," to "do equal right to the poor and to the rich," and to "faithfully and impartially discharge and perform all the duties incumbent upon me ... under the Constitution and the laws of the United States." If we in the judiciary do not have the authority, and indeed the responsibility, to right fundamental wrongs left excused by a majority of the electorate, our whole intricate, constitutional system of checks and balances, as well as the oaths to which we swore, prove to be nothing but shams.

In the coming weeks, Lambert would read Judge Daughtrey's dissent again and again, trying to find a flicker of light.

In Kentucky, Pam Yorksmith was so troubled by the Sixth Circuit's ruling that she started making plans to legally adopt both her sons in South Carolina, which had begun to recognize the marriages of gay couples. The paperwork and legal fees would cost $10,000, but she never again wanted to worry about losing access to her children.

Only weeks earlier, six-month-old Orion Yorksmith had woken up with a terrible cough. "Can you see him struggling to breathe?" The doctor's voice was urgent. "Take off his shirt and look at his ribs."

It was well past midnight, but Yorksmith decided to take the baby straight to the hospital in Cincinnati. She kissed her wife, who would stay home with their older son, Grayden, and buckled Orion into the car. The eleven-mile drive into Ohio seemed endless, and she winced every time she heard Orion wheeze in the backseat.

Yorksmith checked in at the hospital's front desk, holding Orion in her arms. The receptionist called up Orion's original birth records, which listed only Nicole Yorksmith as Orion's mother.

"Are you Georgia?" the receptionist asked, using Nicole's legal first name.

"No. I'm Pam. I'm his other mother."

Yorksmith had the amended birth certificate at home, delivered after Judge Black's ruling in federal court. The state had listed both women's names but added an asterisk underneath with the words, "pursuant to United States District Court." It had felt like yet another insult, but standing there in the hospital waiting room with a wheezing, coughing baby, Yorksmith cursed herself for not thinking to bring it with her.

"You're not listed here," the receptionist said as she scanned Orion's birth records.

"Look, all you have to do is Google my name and you will see that I am this child's other parent."

"I'm sorry," the receptionist replied. "I have to call his mother before we can see him."

Desperate to see a doctor, Yorksmith felt like screaming: I *am* his mother. Instead, with a hospital security guard standing behind her, she said, "You've got to be fucking kidding me."

She carried Orion into the waiting room and texted her wife. "They're going to have to call you before they even look at him."

Thirty minutes passed. She texted again: Have they called yet?

Finally, Yorksmith was shown into a treatment room. The receptionist apologized, but by then, the only thing Yorksmith cared about was seeing a doctor. Driving home on deserted streets sometime after three A.M., with Orion on medication for a bad case of croup, she wondered whether the lawsuit would move forward after the loss in the Sixth Circuit so that, one day, minds and policies would change and no one would question whether she was Orion's "other mom."

It makes you feel like you're playing house in the eyes of other people, she thought, exhausted, as she looked in the rearview mirror at her sleeping son.

Looking back, any number of things could have gone wrong after the loss in the Sixth Circuit as Al and the other lawyers, representing forty-two plaintiffs in four states, scrambled to figure out the best way forward.

Attorneys for the plaintiffs in one state could have opted to appeal to all the judges on the Sixth Circuit, which would have slowed the progression of the case as a whole. Attorneys in another state could have delayed filing an appeal with the U.S. Supreme Court, adding weeks or months to the timeline. But the lawyers in all four states quickly decided to work as a group and appeal together to the Supreme Court. The problem was time.

There were only a handful of months left in the Supreme Court's term. The lawyers studied the court's calendar and, counting backward, discovered that to get the case considered by the end of the court's term in June 2015, they would need to file four legal briefs within days, one for each state. Then, if four of the nine Supreme Court justices wanted to take up the issue, one, some, or all of the cases would move forward to a full-fledged hearing.

Over the next seven days, Al, Susan Sommer at Lambda Legal, and the legal team at the ACLU drew up a lengthy petition that summarized the birth and death certificate cases in Ohio.

In Tennessee, lawyers immediately reached out to Doug Hallward-Driemeier, a former assistant to the U.S. Solicitor General who had argued 14 cases before the Supreme Court. He later became the head of the appellate and Supreme Court division in the Washington, D.C., office of his law firm, Ropes & Gray. A

Harvard Law School graduate, the forty-seven-year-old attorney and Rhodes scholar had always been drawn to the intellectual side of law. In an appeal, the facts of a case were already established and Hallward-Driemeier could scour trial transcripts looking for legal errors and the opportunity to develop new, novel arguments only briefly explored in lower court. He had plenty of experience, having presented appeals before every federal circuit court in the country.

The lawyers from Tennessee wanted Hallward-Driemeier to help draft their petition to the Supreme Court. Though he had never been a lead counsel on a gay rights case, he had drafted a "friend of the court" brief supporting Edie Windsor for a group of progressive organizations that included the Anti-Defamation League.

"Absolutely," he replied when one of the Tennessee lawyers called. "Let's do this."

Early on November 14, eight days after the Sixth Circuit ruling, Al and his team filed their petition with the Supreme Court. Lawyers in Tennessee and Michigan filed their own petitions later that day. On November 18, a legal team that included Shannon Fauver and Dawn Elliott filed a petition for Kentucky.

Jim Obergefell could barely sit still. He created an e-mail file folder called "SCOTUS," short for the Supreme Court of the United States, and posted on Facebook: "I'm not sitting down or shutting up until equality is the law."

He had made a promise to John, and he would see it through right to the end.

20

QUESTION TWO

THE WAITING was misery. On the first day of 2015, after a second New Year's Eve without John, minutes and hours passed in languid succession, and it seemed to Jim as if life had been stripped down to four elusive votes. The weekend came, then Monday and Tuesday, and still, no news.

The nine justices of the U.S. Supreme Court would meet privately to decide whether to hear one or all of the appeals or let the decision by the Sixth Circuit stand, upholding marriage bans in Ohio, Tennessee, Kentucky, and Michigan. For the case to move on, at least four Supreme Court justices had to agree to hear it.

Another day ended with no announcement, and then another week, and by the third Friday in January, Jim could think of little else. He took to Facebook: "It's time for SCOTUS to accept cases and make marriage equality the fact everywhere. I'm tired of waiting." He sat in front of his iPad all afternoon that chilly Friday, trading messages with Joe Vitale in New York and trolling the Internet for news from Washington. Just after three P.M.,

Vitale texted—finally, a decision—and on the Supreme Court's website, Jim saw the most stunning words. Their petitions had been granted. The court had accepted the appeals from all four states and rolled them into one case.. After decades of incremental rulings, the court would consider making same-sex marriage legal in all fifty states.

"Advocates have called same-sex marriage the modern era's most pressing civil rights issue, and the court's action could mark the culmination of an unprecedented upheaval in public opinion and the nation's jurisprudence," journalist Robert Barnes declared in the *Washington Post* that day.

Jim leapt off the couch, his heart pounding. E-mails were pouring in and his cell phone was buzzing, but he stopped to post another message: "Holy hell. I'm going to the U.S. Supreme Court! I miss you, John—this fight is in your honor and memory and as a thank you for almost 21 years of happiness. Going with my friends for their cases, too."

He mentioned Joe Vitale and Rob Talmas, Greg Bourke and Michael De Leon, and Nicole and Pam Yorksmith, who added their own message: "Our family is beyond privileged to have the opportunity to be part of history."

Because Al and his team had been first among the lawyers to file a petition with the Supreme Court, Jim would become the first named plaintiff in the newly consolidated case of *Obergefell v. Hodges*, his name, by virtue of timing, forever linked to the issue of marriage equality, along with that of Richard Hodges, the director of the Ohio Department of Health.

The justices would hear oral arguments in April with a decision expected by the end of the court's session in June.

In Washington, the executive director of the Human Rights Campaign, the country's largest gay rights advocacy group, had been following the case closely. He turned to Fred Sainz and said,

"You need to go out to Cincinnati." Sainz, with wide shoulders and a trim mustache and beard, was the son of Cuban exiles who had organized against Fidel Castro in the 1950s and then fled the country. They settled in bustling Miami, in up-and-coming neighborhoods with mom-and-pop cafés that served up *vaca frita* and shots of Cuban coffee. His mother worked at Woolworth's and his father drove a cab before he got his insurance license and opened his own firm.

At twenty, Sainz landed a job in the George W. Bush White House, the youngest staffer in the West Wing, and went on to jobs running high-profile marketing campaigns. But he was estranged from his father, and by the time Sainz became the communications chief at the Human Rights Campaign in 2010, they hadn't spoken in fifteen years.

If he couldn't move his father, Sainz decided he could try to move others.

Just before the *Windsor* ruling, he had commandeered a conference room, pulled in twenty-five associates, and started thinking about a single image that would represent marriage equality. "Red is for love," someone called out, and the team designed a red-and-pink equal sign that almost overnight went viral on Facebook and Twitter.

Now Sainz needed to meet Jim Obergefell, the named plaintiff in the case that would decide marriage equality, and figure out how Jim might fit into the national campaign. In February, Sainz flew to Cincinnati to meet Jim for breakfast in a hotel café just down the street from the federal courthouse, where Judge Timothy Black had ruled in favor of marriage recognition and the majority of judges on the Sixth Circuit panel had ruled against it.

Sainz ordered a turkey-sausage scramble and looked across the table at the pensive man who had made love and loss seem transcendent, not gay or straight but infinitely human. If Jim became the face of the case, potentially the most important in the history of the gay rights movement, could he withstand the public's glare?

When Fred Sainz was a young boy, his mother once said: *La gente muestra su temple cuando las cosas se ponen difíciles.* People show their mettle when the going gets tough.

Sainz and Jim made small talk through breakfast, chatting about their backgrounds and the chances of a win at the Supreme Court. But over coffee afterward, Jim started telling his story, and in an instant, Sainz knew. Jim described his reaction to the Ohio death certificate during the first meeting with Al, as John lay dying in the bedroom. "It was going to say 'single,'" Jim said. "Even for smart, well-rounded people, sometimes it takes a vivid detail like that to kind of switch the light on."

Sainz immediately decided that they could work together to tell the American public about love and loss in gay families. He invited Jim to Washington to help with HRC's newest campaign, this time to convince voters to sign their name to a "People's Brief" that would be submitted in advance of oral arguments at the Supreme Court.

Back home at HRC headquarters, an eight-story building not far from the White House, Sainz went to work, setting up a website to solicit signatures for the brief. He had only a few days to try to bring in a wide swath of supporters, but in a matter of hours, voters from all fifty states signed their names to the brief. Sainz upped his goal to one hundred thousand signatures and then to two hundred thousand. By the time Jim came to Washington in early March, the brief had grown to 3,500 pages, with so many signatures that it had taken four days of round-the-clock printing to produce the fifty copies required by the Supreme Court.

On a frigid Friday morning, Sainz took Jim to the courthouse to deliver 175,000 pages, stacked in dozens of cardboard boxes. The brief had been written by Roberta Kaplan, a prominent New York civil rights lawyer who had represented Edie Windsor. "Ohio insists that there must be a blank space on Mr. Arthur's death certificate where Mr. Obergefell's name should be," Kaplan wrote. "Not content to deny these men the equal protection of the law in life, it also seeks to deny them dignity even in death."

Jim had woken up that morning feeling different. He had thought about his friend Nanci Vesio, who lost her forty-nine-year-old husband to cancer three months after John died.

"What did you do with all his stuff?" Vesio, a petite mother of two with gray hair and a wide smile, once asked Jim over lunch.

"I got rid of all the sick things first." Jim had decided to paint and rearrange furniture.

"I'm overwhelmed," Vesio said.

"I know. It's okay. Take it day by day."

"Are you lonely?"

"Yeah."

"Does it get any better?"

"Not better," Jim said. "Just different."

On his way to deliver the People's Brief with Fred Sainz, "different" wasn't feeling all that awful. Grief still hovered in the shadows, clawed at his heart so intently that Jim sometimes lost his breath, but he was becoming part of something far larger than himself. Purpose kept the loneliness at bay, and when strangers recognized his face and stopped him to share their stories, Jim felt connected to a tapestry of lives that in some extraordinary way had collided with his own. He had spent his whole life as a private man, but he decided that John's death had given him a public voice, a way to speak up for tens of thousands of gay couples and their families.

Jim had never been to the Supreme Court, and though the grounds were covered with snow, he gaped at the sixteen marble columns and an engraving on top that read EQUAL JUSTICE UNDER

LAW. He climbed the shallow steps to the building's wide oval plaza, passing flagpoles with bronze bases crested with images of the scales and sword, the book, the mask and torch, the pen and mace and the four elements: air, earth, fire, and water. Across the way, the U.S. Capitol building, with its massive white dome, shimmered in the winter sun.

Jim looked around, surveying a scene unlike anything he had thought about when he and John agreed to sue the State of Ohio less than two years earlier. The People's Brief had been signed by 207,551 Americans, an astounding number, and when Jim walked to the doors on the far side of the building to deliver the documents, he wanted more than anything to tell John what they had accomplished.

The nine justices of the Supreme Court had boiled the case down to two questions: whether the Constitution required all fifty states to issue marriage licenses to people of the same sex and whether states with bans should be required to recognize marriages that were legally performed elsewhere.

For each question, only one lawyer could address the court. Al's cases fit into question two.

Years earlier, when Al had an affirmative action case before the Sixth Circuit Court of Appeals, a judge he knew had pulled him aside at a civil rights conference. George Clifton Edwards Jr., nominated to the Sixth Circuit in 1963 by President John F. Kennedy, wasn't involved in Al's case, but he knew Al because Al's law partner had once clerked for the judge.

"Is it going to the Supreme Court?" Judge Edwards asked Al.

"There's a possibility."

"Don't let them take the argument away from you."

"Why do you say that?"

"This is what happens," said the judge, who in 1949 vied to become the mayor of Detroit but lost to another white man after pushing for the equal rights of African Americans. "You have a good lawyer and suddenly they get up to the Supreme Court and they think they can't do it. Don't let that happen."

Al's affirmative action case was never argued at the Supreme Court, but he had been handling his own appeals ever since, mostly because he wanted to keep his clients front and center in appellate debates that could turn dry and academic, focused on legal error rather than the lives of families whose rights were at stake. In all his years as a civil rights lawyer, Al had never appeared before the Supreme Court, and when the lawyers from Kentucky and Tennessee asked if he wanted that chance in *Obergefell*, Al quickly said, "Yes."

It was a career-capping opportunity, but Al wanted to argue for his brother, who thirty years earlier had watched his partner take on the Catholic Diocese of Cleveland and lose, and for the gay community of Cincinnati, which had spent more than a decade living with a city charter that expressly prohibited anti-discrimination laws on the basis of sexual orientation. More than anything, he wanted to argue for John, Jim, and the other plaintiffs in his marriage recognition cases.

He also wanted to argue because he had been told that marriage recognition wasn't the right battle to wage, and now the Supreme Court had decided that the question was important enough to ask it directly. Could marriages simply be dissolved at state lines? Al believed that single question had exposed the impracticality of marriage bans and the serious set of consequences for couples like the Yorksmiths and Jim Obergefell and John Arthur. His plaintiffs had shown the country how much their marriages had meant to them and how, by virtue of geography, their lives had been upended. Surely, Al thought, their stories would nudge the court toward full marriage equality.

Al was hoping his selection for question two would be made through an easy consensus, but the lawyers from Tennessee wanted Doug Hallward-Driemeier to argue instead.

The disagreement was unexpected and awkward: only one lawyer could argue, and two wanted the job. They decided to meet in Louisville at the end of March for another moot court, this time a friendly competition that would pit Al against Hallward-Driemeier, with the best presenter winning the job. It would be a confidential process watched only by the lawyers and the plaintiffs, who would be sworn to secrecy. A videotape of the arguments would be destroyed once the contest was done.

Al could have pushed back, arguing that a newcomer would be no match against a lawyer who had spent the better part of three decades pushing for the rights of the gay community. But he decided that a competitive process would identify the best presenter, and the best presenter would be in the strongest position to help his clients. "We're coming from two very different perspectives and two very different backgrounds, and we should hear how it sounds when we're put into it," Al told his son, Adam.

Al drove to Louisville the day before the moot court, checked into a motel, and stood in front of a mirror, practicing his arguments for hours. The next morning, he walked alone into a conference room on loan from the Louisville Bar Association, his binder tucked into his briefcase. About a hundred people were already there, including Adam, Al's law partner Jennifer Branch, James Esseks from the ACLU, and the lawyers from Tennessee and Kentucky. Jim, Pam and Nicole Yorksmith, and Joe Vitale, who had driven together from Cincinnati, sat near the back of the room.

Hallward-Driemeier and Al flipped a coin to see who would appear first before a panel of three lawyers who would act as Supreme Court justices and then choose an oralist. Al was impressed with the panel, which included Mary Bonauto, the civil rights project director for the Gay & Lesbian Advocates & Defenders, who had helped bring same-sex marriage to Massachusetts and other states.

Al would go first. Hallward-Driemeier left the room and Al began to talk about why marriage recognition bans harmed gay couples from birth to death. The judges interrupted seconds into his twenty-five-minute pitch. Where did the right to marry come from? Why not let the democratic process decide? If you lose on question one—whether there was a constitutional right to marry in all fifty states—don't you also lose on marriage recognition?

Al had expected questions, but about halfway through the argument, he began to feel winded and off track. He worried that he was losing focus and the chance to turn the conversation back to his plaintiffs and their stories. Some of the questions seemed arcane, others too academic. From her seat in the audience, Branch thought it was the hottest bench of judges she had ever seen, flinging questions and often intervening before Al could complete a thought. She was quick to speak up when the panel asked the audience which lawyer did a better job. "Al knows the case," she said. "Al knows the clients."

"He's a mensch," Joe Vitale said, using the Yiddish word for a person of honor. Vitale described Al's frequent phone calls to check in on Cooper, the youngest plaintiff in the case. Was he walking? Was he talking? Did he still like toy cars and *Frozen*?

Jim stood up and said, "Al has earned this. He's been fighting for civil rights for thirty years. He started this case and he deserves it."

"He brought this case," Pam Yorksmith pressed. "He got us here. He should be the one arguing."

Al didn't hear the comments. He was in another room with Hallward-Driemeier watching the video of both arguments. Al thought the other lawyer appeared clear and authoritative, and Al stepped forward when he and Hallward-Driemeier went to see the three judges. "I've got to call this as I see it," Al said. "Doug did a better job."

Al conceded before a vote.

He went to talk to the lawyers waiting in the next room. "I appreciate everybody being here. Based on all the considerations, I think it's best if Doug argues, and he's agreed to do that. I'll remain involved as a member of the team that helps him prepare for this."

But Al was deeply disappointed. He thought about his daughter, Jessica, once a competitive figure skater who practiced for hours for those few critical moments in front of the judges. Sometimes, Al knew, she just missed her jumps.

Al wandered into a brewery with Adam after the moot court, too anxious to drive home to Cincinnati. They settled on a table outside, ordered beers, and watched the Louisville business district shut down for the day. Al was still in the throes of his argument, replaying the questions and answers in his head. Adam had seen it many times before, his father after a day in court, glassy-eyed and distant.

"If I had a couple more days," Al said absently, "I would have been in a different place."

"I was very impressed."

"I don't think I did as well as Doug. He'll do a good job."

Adam was touched by his father's humility. "All of our clients said they trusted you and, in fact, they love you."

At home in Cincinnati, Mimi told Al, "You know you did what's best for your clients."

Tennessee attorney Regina Lambert had missed the moot court because her father was sick, but she heard about Al's concession from one of her cocounsels. "He may never have this opportunity again," Lambert said.

Five days later, she sent Al a note.

> Every single step of the case, from talking with potential plaintiffs to filing a merits brief with the Supremes, has been one amazing learning and growing experience. This 51-year-old lesbian has been forever changed by all.
>
> And yet, you are the one who I cannot get out of my mind for days now. And so, please allow me to tell you: You are a wonderful man, a talented advocate and a gracious team member. You are an inspiration, a mentor and a hero to me. Thank you, Al.

On the Sunday before oral arguments at the Supreme Court, Al went to church.

He knew that the case had galvanized hundreds of thousands of supporters, including groups that had hired their own attorneys to write legal briefs backing the plaintiffs. One had come from 379 employers, including Apple, AT&T, Barnes & Noble, the Coca-Cola Company. "Our successes," the brief read, "depend upon the welfare and morale of all employees, without distinction."

Psychologists, law-enforcement officers, and first responders had weighed in, along with the Anti-Defamation League and the National Association for the Advancement of Colored People. Public officials stepped out by the hundreds, including 211 members of the U.S. House and Senate and 226 U.S. mayors.

The most notable backer was the U.S. government and President Barack Obama, who emphasized in a brief penned by the U.S. solicitor general, "States have burdened petitioners in every aspect of life that marriage touches."

But the opposition was fierce and organized, led by religious groups and a long list of states—Louisiana, Utah, Texas, Alaska,

Arizona, Arkansas, Georgia, Idaho, Kansas, Montana, Nebraska, North Dakota, Oklahoma, South Dakota, West Virginia, South Carolina, and Alabama.

Al had been helping Hallward-Driemeier prepare for oral arguments and had traveled to Nashville and Washington, D.C., for more moot-court sessions. Lawyer Mary Bonauto was also there, since the Gay & Lesbian Advocates & Defenders was working with the Michigan lawyers and Bonauto had been selected to argue question one—whether same-sex marriage should be legal nationwide. Al thought it was a strong, strategic choice. Bonauto was a gay rights pioneer and had worked with Evan Wolfson and others early on to craft the framework for a national marriage road map.

But Al knew that law could be fickle and that decisions could turn sideways even in the face of the most compelling arguments. He had always found a certain comfort in the Sunday worship service at the First Unitarian Church in Cincinnati, with its wood beams and stained-glass window with a picture of the Figure of Truth. Before the congregation that Sunday, Minister Sharon Dittmar presented Al with a gift basket for his journey to the Supreme Court—a flashlight to navigate darkness, contact-lens solution for clarity, scissors to cut through "injustice."

She asked Al to say a few words, and he stood before the group.

"I've been here before and it's scary," he said, thinking back to Issue 3. "In our city, no law could be passed and no board, no entity, could protect gays from discrimination. I thought, 'Wow. That doesn't seem fair. How can we carve out gay people and not protect them?' So we sued. We went to the Sixth Circuit and we lost. Cincinnati became this permanent island of intolerance."

Bith certificate plaintiffs Kelly Noe and Kelly McCracken had come with Al to church, and he looked over at the couple, married in Massachusetts but considered single in Ohio. They held their infant daughter, Ruby. "There is no, in my view, legitimate state interest in harming Ruby and the millions of children like Ruby," Al said. "I'm hopeful that the Supreme Court will see that, but I'm not sure because, like I said, we've been there. With children on the line, a lot of harm can be done, and that is a scary thought.

"That is why your prayers are welcome."

21

OBERGEFELL V. HODGES

PROTESTORS HAD already gathered in front of the Supreme Court when Jim tumbled out of his car in a new tan blazer and raced toward the building, past the young man whipping a Bible above his head, past the old man shouting "God hates fags."

He was late and he knew it, delayed by his third television interview since early that morning. He glanced to his right, where several hundred people hoping to get a ticket to oral arguments waited in a line that had formed the night before and now snaked down the leafy edges of East Capitol Street. In less than twenty-four hours, nine Supreme Court justices would hear Jim's story. Finally, they would know John's name.

For a brief second, Jim felt confident, giddy, wooed by a frenzied crowd that had come to the courthouse because it seemed as if the country was on the cusp of change, the national map tipped toward something close to victory. Volunteers with the ACLU handed out water bottles to the people in line, dripping sweat in the late-April sun. Schoolchildren sang the national anthem and supporters who had driven through the night waved flags above their heads, a canopy of rainbow colors.

All of the plaintiffs were crammed shoulder to shoulder on the courthouse steps except for two-year-old Cooper Talmas-Vitale, the youngest, who toddled from the sidewalk to the plaza and the plaza to the sidewalk in a navy-blue blazer, paying no mind to the photographers who were waiting to snap a final group shot. Jim squeezed in on the end next to David Michener, who had moved his three children from Cincinnati to the Delaware shore after his husband died. Jim adjusted his glasses and smiled.

"We're going to win. I can feel it," someone called from the crowd below.

Jim was always hopeful in the daytime, when friends talked about making history and the hours whizzed by, another reporter needing an interview, another e-mail needing an answer. The restlessness came at night, when thoughts about hearts and minds and the democratic process blurred the tenuous line between faith and uncertainty and Al's words of caution lingered in the dark. *Anything can happen. It could go either way.*

Jim moved away from the steps as reporters called his name. "How do you feel?" "Are you ready for tomorrow?"

He stood before a series of television cameras and clipped a microphone to his lapel. "Standing up for your loved one, for the person you're committed to, it's an honor."

Behind him, an angry man with a meandering gray beard called, "Homosexuals will not enter the Kingdom of God."

"That's all we're asking for," Jim said, "to be treated like any other American citizen."

The bearded man persisted: "God will bring his wrath on this nation."

"Marriage is a fundamental right. I deserve the same as every other American citizen."

"Burn in hell!"

Jim ignored the insults, but it pained him every time, some cold, clueless stranger stomping all over the twenty-one years he had spent with John. He weaved through the crowd and headed across town to a party hosted by Freedom to Marry in a high-rise law office with sweeping views of the Potomac River. The room was crowded with Washington's liberal power brokers and gay rights leaders, including lawyer Evan Wolfson and dozens of couples who had filed lawsuits long before *Obergefell.* Pam and Nicole Yorksmith, mildly exhausted, navigated a double stroller between the tables, with Joe Vitale, Rob Talmas, and Cooper following close behind.

White House Senior Advisor Valerie Jarrett had come to toast the plaintiffs. "The word that really touches me tonight looking at all of you is the word *love*," she said, and the crowd erupted in cheers.

Jim ducked out early, leaving his white wine untouched. In the four days since he had landed in Washington and dropped his bags at a rented townhouse in Georgetown, he had attended three lunches, one brunch, four receptions, two press conferences, and so many one-on-one interviews that he had lost count. His niece, Kara Obergefell, had come in from New York to stay with him, along with old friends Adrienne Cowden and her husband, Eric Avner. They had stocked the townhouse with wine, brie, tortilla chips, toothpaste, and an iron even though Jim's gray suit for his day at the Supreme Court was hanging neatly in a closet upstairs. Avner also brought along a copy of the *Cincinnati Enquirer* with Jim's picture on the front page.

Jim had pointed to his photo and sighed. "I've got a flabby, fat neck."

"We call that a waddle." Avner had made it his mission to keep Jim grounded amid the relentless rounds of media interviews and had recently added two emoticons to his texts, a flame and a pile of poop, so that he could tease Jim about his newly found "hot shit" status.

"It's just a bad angle," Jim's niece said, peering sideways at the picture.

Jim paused for a minute, looking at the newspaper. "It's surreal."

Jim had decided to leave the Freedom to Marry party early because he was throwing a reception for his family and friends, and HRC had offered to host the party at the organization's headquarters. Jim's oldest brother had traveled to town from Sandusky, and John's aunt Paulette had come from Cincinnati. Jim had asked Paulette to go with him to oral arguments because she reminded him of John.

The conference room at HRC was packed when Jim arrived. Fred Sainz was talking to HRC president Chad Griffin, who got his start volunteering for Bill Clinton's presidential campaign and later founded the nonprofit American Foundation for Equal Rights, which successfully challenged the constitutional amendment in California that had banned gay couples from marrying. The attorney general of Virginia and some of HRC's biggest donors were munching on beef skewers. But Jim wandered over to his brother, a construction foreman and the eldest of his five siblings. "I wish John was here," his brother said.

The next morning, Jim woke before dawn. It was shaping up to be an 84-degree Tuesday in Washington when he arrived at the Supreme Court just before seven A.M. for an interview with CNN. After the interview, he found Paulette and hugged her, and together they walked toward their saved spots in a ticket line that stretched down the street and around the corner. The early-morning quiet was pierced by the click, click, click of cameras as Jim and Paulette slipped into place.

"It's official," he whispered. Paulette took his hand and squeezed.

Twice that morning she had called him John, and in the seconds before they were called inside the courthouse, she apologized. "I don't know why I called you that."

"He's just with us today," Jim said.

They entered the Supreme Court through a side door, climbed one flight of stairs, walked down the Great Hall filled with busts of the former chief justices, passed through oak doors, and finally came to the court's chamber, with a forty-four-foot ceiling and twenty-four columns made of marble from Liguria, Italy. Jim could see Al sitting up front with other lawyers.

Outside, a protestor waved a picture of two men having sex on a sign that read FAG MARRIAGE. Inside, in a hushed courtroom at exactly 10:02 A.M., one of the most momentous civil rights cases in nearly half a century began to unfold.

"We'll hear oral argument this morning in Case No. 14-556, *Obergefell v. Hodges* and the consolidated cases."

Obergefell. Jim sucked in his breath when he heard his name.

"Mr. Chief Justice, and may it please the Court." Civil rights lawyer Mary Bonauto would speak first. A petite woman with gray hair and an easy nature, she had been tapped to deliver the argument by the lawyers in Michigan and Kentucky less than four weeks earlier. When she told them that she had never argued before the Supreme Court, one of the lawyers shrugged and shot back, "You've been practicing for this for twenty years." Bonauto, a longtime attorney at Boston's Gay & Lesbian Advocates & Defenders, had been a key player in same-sex marriage wins in Vermont, Massachusetts, and Connecticut and had won the first lawsuits challenging the Defense of Marriage Act.

She had brought along a manila folder with her notes, but as she started to deliver her opening remarks, she decided she didn't need them. "The intimate and committed relationships of same-sex couples, just like those of heterosexual couples, provide mutual support and are the foundation of family life in our society," Bonauto said. "If a legal commitment, responsibility, and protection that is marriage is off-limits to gay people as a class, the stain of unworthiness that follows on individuals and families contravenes the basic constitutional commitment to equal dignity."

The justices quickly interrupted, and Bonauto wasn't surprised. Proceedings in the Supreme Court, like those in all appellate courts, are meant only to determine whether lower courts misinterpreted law. Though lawyers give brief opening remarks, the facts of the case are kept to a minimum in favor of legal analysis and precedent. Judges frequently interrupt with questions, and Bonauto and her team had spent hours trying to anticipate what would be asked and how she would respond. "You'll be lucky to get a word in edgewise," Doug Hallward-Driemeier, the more experienced oralist, had told her.

". . . The argument on the other side is that they're seeking to redefine the institution," Chief Justice John Roberts said. "Every definition that I looked up, prior to about a dozen years ago, defined marriage as unity between a man and a woman as husband and wife. Obviously, if you succeed, that core definition will no longer be operable."

"I hope not, Your Honor," Bonauto said quickly, "because . . . what we're really talking about here is a class of people who are, by state laws, excluded from being able to participate in this institution. And if Your Honor's question is about, Does this really draw a sexual orientation line—"

"No, my question is you're not seeking to *join* the institution. You're seeking to *change* what the institution is. The fundamental core of the institution is the opposite-sex relationship and you want to introduce into it a same-sex relationship."

" . . . If you're talking about the fundamental right to marry as a core male-female institution," Bonauto replied, "I think when we look at the Fourteenth Amendment, we know that it provides enduring guarantees in that what we once viewed as the role of women, or even the role of gay people, is something that has changed in our society."

Justice Anthony Kennedy stepped in, and it seemed to Jim as if the room went still. For weeks, legal scholars and journalists had endlessly speculated that the outcome of the case would rest with the Irish Catholic jurist who had been appointed to the bench by President Reagan but had emerged in recent years as a swing vote in key wins on gay rights. Justice Kennedy had authored the majority opinion in *Windsor* and years earlier had sided with more liberal judges to invalidate criminal sodomy laws. But he had also voted to uphold the right of the Boy Scouts of America to ban gay scoutmasters. Even for the savviest pundits, Justice Kennedy's ideology had been hard to pin down.

In *Obergefell*, it seemed likely the court would split, with left-leaning Justices Ruth Bader Ginsburg, Sonia Sotomayor, Elena Kagan, and Stephen Breyer siding with the plaintiffs and more conservative justices Clarence Thomas, John Roberts, Samuel Alito, and Antonin Scalia backing the states. The unknown was Kennedy, who was pressing Bonauto about the traditional definition of marriage. " . . . This definition has been with us for millennia. And it's very difficult for the court to say, 'Oh, well, we . . . we know better,' " he said.

"Well, I don't think this is a question of the court knowing better," Bonauto replied. "When we think about the debate, the place of gay people in our civic society is something that has been contested for more than a century. . . . It was over twenty years ago that the Hawaii Supreme Court seemed to indicate that it would rule in favor of marriage, and the American people have been debating and discussing this. It has been exhaustively aired, and the bottom line is that gay and lesbian families live in communities as neighbors throughout this whole country. And people have seen this—"

"You argue in your brief that the primary purpose of the Michigan law limiting marriage to a man and a woman was to demean gay people. Is that correct?" asked Justice Alito, nominated by George W. Bush in 2005 and considered one of the court's most conservative thinkers.

"The Michigan statute and amendment certainly went out of their way to say that gay people were in some sense antithetical to the good of society," she replied. "They wrote that—"

"And did you say in your brief that the primary purpose of that was to demean gay people?" Alito asked.

"I think it has that effect, Your Honor. I do. Now, at the same time—"

"Is that true just in Michigan or is that true of every other state that has a similar definition of marriage?"

"Well, if we're talking about the states that have constitutional amendments, many of them are similar. . . . But even if there's not a purpose to demean, I think the . . . commonality among all of the statutes, whether they were enacted long ago or more recently, is that they encompass moral judgments and stereotypes about gay people. Even if you think about something one hundred years ago, gay people were not worthy of the concern of the government. . . ."

"Well, how do you account for the fact that, as far as I'm aware, until the end of the twentieth century, there never was a nation or a culture that recognized marriage between two people of the same sex?" Justice Alito asked. "Now, we can infer from that that those nations and those cultures all thought that there was some rational, practical purpose for defining marriage in that way, or is it your argument that they were all operating independently based solely on irrational stereotypes and prejudice?"

"Your Honor, my position is that times can blind," Bonauto replied, and pointed to the issue of discrimination based on gender.

"And if you think about the example of sex discrimination . . . I assume it was protected by the Fourteenth Amendment, but it took over one hundred years for this court to recognize that a sex classification contravened the Constitution. . . ."

"I don't really think you answered my question," Justice Alito said.

"I'm sorry."

"Can we infer that these societies all thought that there was a rational reason for this and a practical reason for this?"

"I don't know what other societies assumed, but I do believe that times can blind and it takes time to see stereotypes and to see the common humanity of people who had once been ignored or excluded."

From his seat near the front of the courtroom, Al was fixed on Justice Kennedy, trying to read his facial expressions and reactions to the questions and answers. So far, none of the questions were particularly surprising, and Al thought Bonauto appeared poised, on point, and decisive.

"Do you know of any society, prior to the Netherlands in 2001, that permitted same-sex marriage?"

The question came from Justice Scalia, the longest-serving jurist on the court and among the most vocal.

"As a legal matter, Your Honor?" Bonauto replied.

"As a legal matter."

"I [do] not."

"For millennia, not a single other society until the Netherlands in 2001, and you're telling me that they were all—I don't know. What?"

"No. What I'm saying is taking that tradition as it is, one still needs, the court still needs a reason to maintain that tradition when it has the effect—"

"Well, the issue, of course, is not whether there should be same-sex marriage, but who should decide the point. . . . You're asking us to decide it for this society when no other society until 2001 ever had it?" Scalia asked.

Al was taken aback by the question. What matters here, he thought, is not what other countries had done but what was fair and lawful in the United States, bound by a Constitution that protected the rights of all its citizens.

". . . Did they have same-sex marriage in ancient Greece?" Justice Alito asked a few minutes later.

Now we're talking about the definition of marriage from before the Middle Ages? Al looked from the judge to Bonauto, kicking himself for not anticipating that kind of question at the moot-court sessions they had held before the argument. But Bonauto had studied the classics in college, and though she believed the question was off base, she answered easily.

"I don't believe they had anything comparable to what we have, Your Honor. You know, and we're talking about—"

"Well, they had marriage, didn't they?" Justice Alito asked.

"Yeah, they had—yes. They had some sort of marriage."

"And they had same-sex relations, did they not?"

"Yes. And they also were able to—"

"People like Plato wrote in favor of that, did he not?"

"In favor of?"

"Same-sex—wrote approvingly of same-sex relationships, did he not?"

"I believe so, Your Honor."

"So their limiting marriage to couples of the opposite sex was not based on prejudice against gay people, was it?"

"I can't speak to what was happening with the ancient philosophers—" Bonauto replied.

Justice Kennedy cut in. "But it's you, you said that, 'Well, marriage is different because it's controlled by the government.' But from a historical, from [an] anthropological standpoint, Justice Scalia was very careful to talk about societies. Justice Alito talked about cultures. If you read about . . . ancient peoples, they didn't have a government like this. They made it themselves and it was a man and a woman."

"There were certainly marriages . . . prior to the United States forming, and we recognize that," Bonauto replied. "But when our nation did form into this union in 1787 and then when it affirmed the Fourteenth Amendment in 1868, that's when we made—our nation collectively made a commitment to individual liberty and equality."

Later, Chief Justice Roberts said, "One of the things that's truly extraordinary about this whole issue is how quickly has been the acceptance of your position across broad elements of society. . . . But if you prevail here, there will be no more debate. I mean, closing the debate can close minds, and it will have a consequence on how this new institution is accepted. People feel differently about something if they have a chance to vote on it than if it's imposed on them by the courts."

" . . . When I think about acceptance," Bonauto said, "I think about the nation as a whole, and there are places where, again, there are no protections, virtually no protections for gay and lesbian people in employment, in parenting. You know, the Michigan petitioners, for example, are not allowed to be parents of their own children, the children that the State of Michigan has placed with them and approved of their adoptions."

"Miss Bonauto, I'm concerned about the wisdom of this court imposing through the Constitution a requirement of action which is unpalatable to many of our citizens for religious reasons," Justice

Scalia said. "They are not likely to change their view about what marriage consists of. And were the states to adopt it by law, they could make exceptions to what is required for same-sex marriage, who has to honor it and so forth . . ."

"Your Honor, of course the Constitution will continue to apply, and right to this day, no clergy is forced to marry any couple that they don't want to marry. We have those protections."

"But right to this day," Justice Scalia continued, "we have never held that there is a constitutional right for these two people to marry, and the minister is—to the extent he's conducting a civil marriage—he's an instrument of the state. I don't see how you could possibly allow that minister to say, 'I will only marry a man and a woman. I will not marry two men.' Which means you would, you could, have ministers who conduct real marriages that are civilly enforceable at the National Cathedral, but at St. Matthew's downtown because that minister refuses to marry two men, and therefore, cannot be given the state power to make a real state marriage. I don't see any answer to that. I really don't."

Jim twisted in his seat. Though spectators had to leave their wallets, keys, and cell phones in stacked rows of lockers just outside the courtroom, Jim had decided to carry in a few sheets of paper and a pen to take notes. He paused and studied Justice Scalia, stung by the comment. Jim looked down at his lap, unsure what to write next.

22

STATE INTEREST

MICHIGAN ATTORNEY John Bursch, an experienced Supreme Court litigator, walked quickly to the lectern and faced the nine justices.

"This case isn't about how to define marriage," said Bursch, the former solicitor general of Michigan. "It's about who gets to decide that question. Is it the people acting through the democratic process or is it the federal courts? And we're asking you to affirm every individual's fundamental liberty interest in deciding the meaning of marriage. . . ."

"I'm sorry," interjected Justice Sotomayor, the first justice of Hispanic heritage and the third woman to serve on the court. "Nobody is taking that away from anybody. Every single individual in this society chooses, if they can, their sexual orientation or who to marry or not marry. I suspect even with us giving gays rights to marry that there's some gay people who will choose not to, just as there's some heterosexual couples who choose not to marry. So we're not taking anybody's liberty away."

"But we're talking about the fundamental liberty interest in deciding the question of what marriage means, and to get that—"

" . . . I thought that I heard the answer to the question," Justice Breyer cut in. " . . . What I heard was, one, marriage is fundamental. I mean, certainly that's true for ten thousand years. And marriage, as the states administer it, is open to vast numbers of people who both have children, adopt children, don't have children, all over the place. But there is one group of people whom they won't open marriage to so they have no possibility to participate in that fundamental liberty. That is people of the same sex who wish to marry."

He went on, "And so we ask, 'Why?' And the answer we get is, 'Well, people have always done it.' You know, you could have answered that one the same way we talk about racial segregation. Or two, because certain religious groups do think it's a sin, and I believe they sincerely think it. There's no question about their sincerity, but is a purely religious reason on some part of some people sufficient? And then when I look for reasons three, four, and five, I don't find them. . . ."

"Justice Breyer," Bursch replied, "those answers one and two are not our answers."

"Good."

"Our answer number one is that the marriage institution did not develop to deny dignity or to give second-class status to anyone. It developed to serve purposes that, by their nature, arrive from biology. Now imagine a world today where we had no marriage at all. Men and women would still be getting together and creating children, but they wouldn't be attached to each other in any social institution. . . ."

"Mr. Bursch, I understand that argument," Justice Kagan, the fourth woman to serve on the court, said. "It's the principal argument that you make in your briefs, that same-sex marriage doesn't advance this state interest in regulating procreation. Let's just assume for the moment that's so. Obviously, same-sex partners cannot procreate themselves. But . . . are you saying that recognizing same-sex marriage will impinge upon that state interest, will harm that state interest in regulating procreation through marriage?"

"We are saying that, Your Honor . . . "

"How could that be?" Justice Ginsburg, the second woman on the court, asked, "because all of the incentives, all of the benefits that marriage affords would still be available. So you're not taking away anything from heterosexual couples. They would have the very same incentive to marry, all the benefits that come with marriage that they do now."

"Justice Kagan and Justice Ginsburg," Bursch said, "it has to do with the societal understanding of what marriage means. This is a much bigger idea than any particular couple and what a marriage might mean to them or to their children. And when you change the definition of marriage to delink the idea that we're binding children with their biological mom and dad, that has consequences."

Al thought about Joe Vitale and Rob Talmas, Pam and Nicole Yorksmith, and all the plaintiffs' children who were waiting in the Supreme Court's cafeteria for their parents.

"That's the problem," Justice Sotomayor said.

"If I could—"

"Marriage doesn't do that on any level. [In] how many married couples do fathers . . . walk away from their children?"

"Justice—"

"So it's not that the institution alone does it and that without it, that father is going to stay in the marriage. He made a choice."

"Justice—"

" . . . How does withholding marriage from one group, same-sex couples, increase the value to the other group?"

"Justice Sotomayor, there's all kinds of societal pressures that are already delinking that reason that the state advances for marriage, keeping kids and their biological moms and dads together whenever possible."

"I think," Justice Kagan said moments later, "before something as fundamental to a society and to individuals as marriage, before an exclusion of this kind can be made in that institution, the state needs some reason for that exclusion."

". . . Well, first, it wasn't a reason for an exclusion," Bursch replied. "It was a definition to solve a particular problem. But the reason why there's harm if you change the definition because, in people's minds, if marriage and creating children don't have anything to do with each other, then what do you expect? You expect more children out of marriage."

"But do you think that that's what it would do, Mr. Bursch, that if one allowed same-sex marriage, one would be announcing to the world that marriage and children have nothing to do with each other?" Justice Kagan asked.

"Not in the abstract, Your Honor. That kind of example—"

"Well, not in the abstract, not in the concrete."

Justice Kennedy cut in seconds later. "You argued in your brief, and Justice Kagan was quite correct to say, that you're saying that this harms conventional marriage. . . ."

"Justice Kennedy, to be perfectly clear, the state of Michigan values the dignity and worth of every human being, no matter their orientation or how they choose to live their life," Bursch said. "That's not what this case is about. Our point is that when you change something as fundamental as the marriage definition, as Chief Justice Roberts was saying, the dictionary definition which has existed for millennia, and you apply that over generations, that those changes matter."

". . . You know," Justice Sotomayor said moments later, "the problem is that I don't actually accept your starting premise. The right to marriage is, I think, embedded in our constitutional law. It is a fundamental right. We've said it in a number of cases. The issue is you can't narrow it down to say, 'But is gay marriage fundamental?' "

Justice Kagan added, "See, to me it seems as though you are doing something very different that we've never done before which is you are defining constitutional rights in terms of the kinds of people that can exercise them."

Since the courtroom was packed, Greg Bourke, Michael De Leon, Isaiah, and Bella only had tickets inside for question two, and so they waited in the heat at the front of the building, clutching a small Kentucky state flag. "We should be viewed as an actual family and not hated," Isaiah told a reporter as he glanced at his fathers, "for what they are."

Protesters were squeezed on the sidewalks, on the grass, in the streets, but there was jubilation, too, as couples kissed and chanted and the Gay Men's Chorus sang under a tree. The entire fifth-grade class from Georgetown Day School watched in bright-yellow T-shirts with rainbow stickers, and nearby, a priest told a small group, "It's important to see that Christians have a heart."

Finally, Bourke and De Leon were called inside, and they quickly kissed Bella and Isaiah, who went to wait in the cafeteria. "I feel like I am literally shaking on the inside," Bourke told another father from Kentucky as they stood in a stuffy holding area just outside the courtroom. Inside, sitting next to Pam and Nicole Yorksmith, Bourke took long, slow breaths and reached for his husband's hand.

At the front of the courtroom, Doug Hallward-Driemeier sat at the counsel's table with Al, attorney Dawn Elliott from Kentucky, and Abby Rubenfeld, who had led the marriage recognition case in Tennessee. U.S. Solicitor General Donald B Verrilli Jr., who had come to support the plaintiffs on behalf of the president, sat nearby.

At 11:39 A.M., Hallward-Driemeier stood up to address the court.

Al tensed. It was finally time. He had always wanted an outright win for marriage equality, but he needed a win for his clients no matter the outcome of question one. This second question and the plaintiffs who were posing it, Al believed, would move the court to at least allow marriages to travel across state lines. But he also hoped his case had exposed the tremendous hardships faced by gay couples in birth and in death and that, in the end, their stories would convince the court that same-sex marriage in all fifty states couldn't wait on the democratic process.

"Mr. Chief Justice, and may it please the court," Hallward-Driemeier started. "The question two petitioners are already married. They have established those enduring relationships and they have a liberty interest that is of fundamental importance to these couples and their children. A state should not be allowed to effectively dissolve that marriage without a sufficiently important justification to do so. . . . "

". . . Let's say someone gets married in a country that permits polygamy. Does a state have to acknowledge that marriage?" Justice Scalia asked moments later.

"Well, of course, the state could assert justifications for not doing so, and I think there would be justifications."

"What would the justification be? That it's contrary to the state's public policy, I presume. Right?"

"Well, no, Your Honor. I think that the justification would be that the state doesn't have such an institution. A polygamous relationship would raise all kinds of questions that the state's marriage laws don't address."

"What if it's not a plural relationship?" Justice Alito asked. "What if one state says that individuals can marry at the age of puberty. So a twelve-year-old female can marry. Would another state be obligated to recognize that marriage?"

"I think probably not," Hallward-Driemeier said. "But the state would have, in that instance, a sufficiently important interest in protecting the true consent of the married person. . . . "

"I think your argument is pretty much the exact opposite of the argument of the petitioners in the prior case," Chief Justice Roberts said about halfway through the hearing. "The argument that was presented against them is: you can't do this, we've never done this before, recognized same-sex marriage. And now you're saying, 'Well, they can't *not* recognize same-sex marriages because they've never not recognized marriages before that were lawfully performed in other states."

"Well, what—"

"You've got to decide one or the other if you win."

Hallward-Driemeier paused. He had prepared for this, and there was no hesitation when he said, "No, I don't think so at all, Your Honor. And I think that what's essential and common between us is that we recognize that the marriage that our petitioners have entered into is a marriage. It is that same institution, that same most important relationship of one's life that this court has held out as fundamental. . . ."

"It's our clients who take marriage seriously," he went on moments later. "They took vows to each other and bought into an institution that, indeed, as this court has said, predates the Bill of Rights, that is the most important and fundamental in their lives, and the state should offer something more than mere pretext as ground to destroy it."

Al had been hoping that Hallward-Driemeier would personalize the arguments somehow, describe the plaintiffs who had put their lives on hold and stepped out publicly to defend their marriages and protect their children. Al shifted forward in his seat when that happened at the close of the hearing.

Hallward-Driemeier mentioned Val Tanco, Sophy Jesty, and their daughter, Emi, who was waiting outside with Jesty's twin sister. ". . . Tennessee would treat Dr. Jesty not as a mom, but as a legal stranger with no right to visit her child, no right to make medical decisions for her. These laws have real import for real people."

Hallward-Driemeier turned to the plaintiffs of Ohio. "Even Jim Obergefell's husband's death certificate will not reflect the fact that he was married or the name of his husband. The state has no legitimate interest for denying them the dignity of that last fact regarding his life."

Tanco, Jesty, and Jim cried softly. Their stories had been told out loud in the official record of the historic hearing.

Hallward-Driemeier straightened his shoulders and pleaded with the court, "I urge the court not to enshrine in our Constitution a second-class status of these petitioners' marriages."

Hallward-Driemeier had appeared before the Supreme Court many times and had never shed a tear, but when he walked to the counsel's table, his eyes were moist.

The plaintiffs and lawyers spilled out of the courtroom when it was over, down the Great Hall, through the heavy bronze doors, out onto the steep flight of stairs to the plaza below, Bourke and De Leon, Talmas and Vitale, Pam and Nicole Yorksmith. Sophy Jesty, arm in arm with her wife and attorney Regina Lambert, could feel the roar from the crowd roll up the steps. Jim's face was red from crying, and he stopped, high above Washington, and put his hand over his heart. The cheers grew louder. *Do you hear that, John?*

Dozens of journalists were waiting in a thick tangle of cameras set up in a semicircle near the fountain on his left, but still Jim stood, gazing across the Capitol. Al gripped Jim's shoulders, rubbed his back. "I'm proud of you," Al whispered, and together they descended the steps.

Bourke moved to the cameras, his voice barely audible over the cheers. "They saw our families. They saw our love."

Jim said, "I trust in the Supreme Court. . . . We all deserve the same civil rights, the same fundamental rights."

Jim ducked into a car with Al and headed to a luncheon at the ACLU's offices downtown. Already Jim had twenty-eight new texts, thirty unread e-mails, forty-three Facebook friend requests, and 140 Facebook messages. But his friends and family were waiting at the lunch, some wearing pink TEAM HALF FULL T-shirts in John's honor. James Esseks, Al's cocounsel and the head of the ACLU's LGBT project, raised a glass of champagne. "We needed people to step forward . . . to share their stories, their families, their heartbreak."

Esseks choked up. "Cheers!"

In good times, Al's father had always shouted, "Hip, hip, hooray!" Standing next to Esseks, Al cried, "Hip, hip . . ."

"Hooray!" the room erupted.

Jim turned around, to Paulette, to Eric Avner and Adrienne Cowden, to his niece, to his brother, to other friends and family members. They held up their glasses and said, in unison, "To John."

23

SUNLIGHT

SIX DAYS later, Al went for a bike ride.

He pedaled out of Cincinnati with his daughter, Jessica, on a tandem bicycle fixed with a sack of gear. They rode west, across a creek, along roads made of gravel, to a general store in Indiana that served fried-bologna sandwiches. They rode on to Illinois, where the wind along the side of the highway nearly knocked them to the ground, and near a town called Salem, hot pebbles from street tar whipped at their knees.

They reached St. Louis, sunburned and aching, and spent the day talking to civil rights leaders about the violence in nearby Ferguson and how, in Cincinnati, Al had worked with the city on reforms that brought fairness and accountability to policing. They left in driving rain, their faces freckled with mud, but thirty-eight miles of rolling countryside was ahead and then the banks of the Mississippi River, where they stopped, awestruck.

They rolled into Memphis and ate barbecue with the Tennessee marriage equality team. They crossed deeper into the South, where backyards had barbecue smokers and small towns no longer had downtowns, save for a Baptist church. They went to a Sunday service in a building made of white clapboard and watched the congregation of thirteen throw coins and cash into a collection bucket, raising $504 for the poor. They biked on, through rain that struck sideways, and when they couldn't find anyplace to eat in the heart of the Mississippi Delta, a family of Indian, Bangladeshi, and African American descent invited them to a graduation party for a defensive tackle named BJ who was leaving for college.

They wound through the windy back roads of Port Allen, Louisiana, past chemical plants and levees, and when they got caught on a bridge they couldn't cross on their two-person bike, a man in a pickup truck heaved it onto his trailer and shuttled Al and Jessica to the other side. Finally, with 1,360 miles behind them, a blog filled with accounts of their travels, and $27,000 raised from their charity ride, which they would donate to criminal justice reform, they met up with Mimi in New Orleans.

Al had always taken to the road to clear his head when a case settled or a trial ended, pedaling his steel-frame bike around Lake Erie, Wisconsin, Illinois, and once, just before he met Jim Obergefell and John Arthur, 1,500 miles from El Paso to San Francisco to see his grandson on his first birthday. After months of grueling legal work, the physical exertion was comforting, and when he arrived in New Orleans in late May, a month after oral arguments at the Supreme Court, Al felt balanced, serene. He had brought a radio splitter on the trip to replay the oral arguments for Jessica while they biked, but Al knew there was nothing more he could do except wait for a ruling.

Jim decided to fly to Washington as the possible decision drew closer. The Supreme Court had scheduled a series of "decision days" before its session ended in late June, but only the nine justices knew for sure which rulings would come down when. Though the court often saved its biggest decisions for the end of the term, Jim wanted to be at the courthouse for every decision day in the second half of June, just in case.

The first was on a Monday. He arrived at the Supreme Court just after the sun came up and glanced at the small, scattered crowd, which included a lone gray-haired woman stationed at the foot of the steps bearing a sign that read WARNING! GOD DRAWS A LINE ON GAY MARRIAGE. He took a spot in line at the front of the courthouse, wondering whether the justices would surprise everyone and announce a decision on this still morning, when Washington was waking up to another summer workweek.

He was given a bright-orange admissions ticket and directed through the doors on the far-left side of the building, down the Great Hall and into the courtroom. He rose when the clerk called the hearing to order. He sat after the nine Supreme Court justices were seated. He waited, breathing slowly, as each new decision was announced, one on an immigration case and another on a bankruptcy case. There was no ruling on same-sex marriage.

He returned on Thursday, passing the gray-haired woman with her sign, and then again the following Monday, this time carrying a picture of John in his breast pocket. A massive bank of television cameras had set up at the steps of the courthouse, but still no decision.

It seemed to Jim as if he had spent the better part of two years waiting, grief and indignation fueling a journey that would finally end here, inside the marbled halls of the Supreme Court. The tension mounted as each day ended without an answer, and he found himself whispering to John late into the night. *Soon.*

On Thursday, June 25, the crowds were back, clogging the sidewalks and weaving around police dogs and news crews. Someone brought red helium balloons that spelled LOVE, and tourists stopped to snap pictures. There were only three days left in the court's session and decisions on seven cases were still to come, including a ruling on President Obama's health care law and another on a housing discrimination case out of Texas. For the fourth time, Jim clutched his orange admissions ticket and walked through the doors on the left side of the building, down the Great Hall, and into the courtroom.

Outside at ten A.M., the crowd quieted, waiting on word from the journalists inside. Thirteen years earlier, two Washington lawyers had started a blog that tracked the happenings of the Supreme Court, and at 10:01 A.M., SCOTUSblog posted: "We have the first opinion."

Television station interns who had been waiting in the court's pressroom came running across the expanse of the plaza minutes later to deliver paper copies of the ruling to the journalists outside. This one was on the Texas housing discrimination case.

About 10:10 A.M., the interns raced across the plaza to deliver the second opinion, this one upholding the Affordable Care Act, a decisive win for the president. Another minute passed, then two. Again the crowd waited. But bloggers inside the courthouse announced that there would be no more opinions that day.

Jim walked outside a few minutes later. Five more decisions were left, spread out over Friday and the following Monday, the last day of the court's term. Friday was a particularly significant day in the history of gay rights because two years earlier on June 26, the court had announced the *Windsor* decision.

The court had never been one to stand on sentiment, but in Cincinnati, Adam Gerhardstein urged his father to catch the next flight to Washington. Al ran home to pack a bag, tossing in his favorite blue tie, given to him by his late father-in-law, the judge. Al would wear it to the Supreme Court. That night, Al flew to Washington, two days earlier than planned, just in case.

It was still dark outside on Friday morning when Jim fastened an American flag pin to his lapel and straightened his lavender-and-white bow tie. He was tense and his mind was racing, thinking about the days spent waiting and the ruling that could come within hours. He pulled his key chain from the dresser and stashed it in his pocket. It was a gold cube with good-luck symbols on all sides, a gift from a friend who had brought it home from Ireland, and Jim had taken to carrying it with him to the Supreme Court. Finally, he slipped on his wedding ring, with John's ashes encased inside.

Outside there was a cool breeze, a welcome change to the muggy heat that had gripped Washington for days. For the fifth time that month, Jim rode across town to the Supreme Court, where a line of people hoping to get inside once again stood in a line that spanned the length of the street and round the corner. Jim found Al in front of the court building and the two men embraced as supporters waved signs: AMERICA IS READY.

"I'm nervous," Al told Jim.

"That sounds familiar," Jim said, smiling.

"Today may be the day."

Jim paused. "I'm so glad you're here."

Pam and Nicole Yorksmith hopped out of a car with their double stroller, and it felt like something of a reunion to Jim as he watched Al bend down to say hello to four-year-old Grayden, who was sporting a bow tie.

"How are you feeling?" Nicole Yorksmith asked Jim as Grayden darted up the steps to explore the fountain on the plaza.

"Nervous as hell."

Greg Bourke, Michael De Leon, Bella, and Isaiah had come from Kentucky, and they stood in line behind the Yorksmiths. Jim was hoping to see Joe Vitale and Rob Talmas, but they were home in New York, waiting in the greenroom of MSNBC Studios for a morning interview. The two fathers had started building a legacy box for Cooper, filled with newspaper clips and pictures of the Supreme Court.

Jim was first in line again, and this time, the admissions tickets for the general public were lavender, a color often used in connection with the gay community. *Maybe it's a sign, John.* Jim led the line inside the building, down the Great Hall, through security, and into the courtroom. He sat in the center section, and when he leaned forward, he could see Al up front, along with Mary Bonauto, Doug Hallward-Driemeier, James Esseks, and Susan Sommer. What would have happened, Jim thought, if Al had decided the case wasn't winnable?

At 9:55 A.M., a buzzer sounded, a five-minute warning that court was about to begin. The room was utterly still. Jim breathed deeply and waited. The day before, SCOTUSblog had logged 1.1 million hits. Editor Amy Howe, writing about the case, called it "One of the biggest decisions in recent history."

A second buzzer sounded, and the nine justices filed in. Jim expected to hear the name of another case, just as he had during each of his four prior visits, but this time he heard his own, along with the case number he had memorized, and the words were nothing less than stunning. For a split second, it seemed the only thing he could hear was his heart pounding as Chief Justice John Roberts said, "Justice Kennedy has our opinion this morning in Case 14-556, *Obergefell v. Hodges* and the consolidated cases."

Three seats away from Jim, Nicole Yorksmith squeezed her wife's hand. Bourke and De Leon inched closer to their children.

Justice Kennedy's voice was soft, thoughtful. "Since the dawn of history, marriage has transformed strangers into relatives. This binds families and societies together, and it must be acknowledged that the opposite-sex character of marriage, one man, one woman, has long been viewed as essential to its very nature and purpose.

And the Court's analysis and the opinion today begins with these millennia of human experience, but it does not end there. For the history of marriage is one of both continuity and change. . . ."

Pam Yorksmith was nodding.

"Until recent decades, few persons had even thought about or considered the concept of same-sex marriage. In part, that is because homosexuality was condemned and criminalized by many states through the mid-twentieth century. It was deemed an illness by most experts. Of necessity, truthful declaration by same-sex couples of what was in their hearts had to remain unspoken . . ."

"Although the limitation of marriage to opposite-sex couples may long have seemed natural and just, its inconsistency with the right to marry is now manifest. It would diminish the personhood of same-sex couples to deny them this liberty. . . ."

". . . The court now holds that same-sex couples may exercise the fundamental right to marry in all states. No longer may this liberty be denied to them."

Jim was so moved that he sobbed quietly. Pam Yorksmith held her hand over her mouth. Isaiah looked to his fathers and mouthed, "We won?" Michael De Leon nodded emphatically. Before the hearing began, Al had described the defeat in the Issue 3 case to a White House lawyer sitting next to him. Now, his face wet, Al passed a note to the lawyer. "I guess I'll get over it now." All around him, seasoned Supreme Court lawyers and advocates cried quietly. Susan Sommer felt light-headed and realized that she had been holding her breath.

"No union is more profound than marriage," Justice Kennedy went on, "for it embodies the highest ideals of love, fidelity, devotion, sacrifice, and family. In forming a marital union, two people become something greater than they once were, and it would misunderstand petitioners to say that they disrespect or diminish the idea of marriage in these cases. They are pleased that they do respect it, but they respect it so deeply they seek to find its fulfillment for themselves. They ask for equal dignity in the eyes of the law and the Constitution grants them that right."

In Tennessee that morning, Val Tanco had woken up with tingling hands and decided it was a sign that something significant was about to happen. Later, Tanco, Jesty, Lambert, and fifteen-month-old Emi huddled together outside the echocardiogram office at the University of Tennessee's vet school, scanning Lambert's iPad for news of a decision.

At 10:01 A.M., SCOTUSblog posted: Marriage.

Lambert shouted, "Marriage!"

The blog posted: Kennedy has the decision.

Lambert shouted, "Kennedy!"

They knew they had won. Lambert's hands started shaking, but all through that day, through celebrations and press conferences, she would maintain her composure. Only later would she remember the nights she had spent as a young girl, begging God not to make her gay. In Justice Kennedy's decision, Regina Lambert found dignity.

In New York, Joe Vitale and Rob Talmas were still in the greenroom of MSNBC when a production assistant barreled in and said, "Marriage ruling in—you won—need you two in the studio now." At home, Cooper's uncle shouted, "Daddies won!" Cooper, who was watching cartoons, threw his arms in the air.

In Kentucky, lawyers Shannon Fauver and Dawn Elliott, who had never filed a civil rights lawsuit and had now won a case that changed the law of the land, popped champagne and then raced to the courthouse to watch couples exchange vows.

In Michigan, lawyer Carole Stanyar danced and cheered with her legal team and plaintiffs until midnight in a packed Ann Arbor courtyard, where couples married under a makeshift archway.

In Cincinnati, Judge Black was celebrating his wedding anniversary. He posted on Instagram: "Marnie and I married 39 years ago today. And today, as we celebrate our love, we welcome all to join us in the glory of marriage."

In Al Gerhardstein's law office, Jennifer Branch, Adam Gerhardstein, the rest of the legal team, and several clients ate wedding cake. Scott Knox stopped by to celebrate and then went over to the courthouse to watch couples get married. A week earlier, Knox had contacted the probate judge to make sure the staff at the courthouse was prepared to issue marriage licenses. In a town once considered among the most antigay in the country, Republican judge Ralph Winkler asked Knox, "Can you get me a list of same-sex couples that want to get married, along with ministers and judges, so that I can give it out to my staff? This will be the biggest day in their lives, and I don't want my staff just doing their jobs. I want them happy and enthusiastic about it."

In rural Ohio, John's aunt Paulette was driving into town with her grandson. Her cell phone rang. "I just want to hug you," said Mike Thomas, a friend who had directed Paulette in a community theater production of *Steel Magnolias* the month that John had died.

"Did something happen with the Supreme Court?" she asked.

"There was a decision."

"Which way did it go?" Paulette gripped the steering wheel.

"Our way."

Later, she told Jim, "This is the ultimate legacy you could have given to John."

At the headquarters of Freedom to Marry in New York, Evan Wolfson, who had first called for marriage equality in his 1983 thesis at Harvard Law School, whispered, "Oh my God. We won," when his Twitter feed broke the news. "That only took thirty-two years."

Inside the Supreme Court, Al and Jim could hear the cheering, chanting crowd even through the courthouse's bronze doors, seventeen feet high and nine feet wide. The frenzied celebration would be captured by newspapers and television stations worldwide.

Later, Al would call Mimi and the mayor of Cincinnati, who would marry five gay couples that evening on Fountain Square. Jim would take a call from the president of the United States. In courthouses across the country, same-sex couples would exchange wedding vows, and in Washington, the White House would be lit in rainbow colors.

But for a minute or two, in a dark corner of the Great Hall, there was only Jim and Al. "This is amazing," Al said softly, putting a hand on Jim's shoulder, "just amazing."

Jim was nodding, trying to collect his thoughts, grateful to Al, missing John. *Baby, can you believe this?* Then they were pushed toward the side doors of the courthouse, the same way they had come in an hour earlier, when the world looked entirely different, and a moment later, Jim and Al stepped together into the blazing summer sun.

EPILOGUE

WHEN DUSK came on decision day, a rainbow streaked the sky above Cincinnati. Al Gerhardstein saw it just as his flight from Washington, D.C., touched down in Ohio, and he smiled all the way to a local restaurant, where he embraced his son, raised a glass, and toasted, "To love."

The next day, Cincinnati threw its annual gay pride parade, coincidentally scheduled for June 27. When Issue 3 became law, Al was so furious with Cincinnati that he considered leaving town. But now the town wanted to welcome him home. All along the sidewalks of the city's business district, supporters cheered and kissed and cried, waving to Al, Jim, and Paulette Roberts as they moved along the parade route in a 1962 Lincoln convertible driven by funeral home director Robert Grunn. Pam and Nicole Yorksmith walked alongside the car with a rainbow flag and Orion in a T-shirt that declared SCOTUS BABY.

The win at the Supreme Court was the sweetest victory in Al's legal career. He had based his life's work on the Constitution and the Constitution had done its job, protecting people like John Arthur, who wanted only to die a married man. Al was profoundly grateful. Soon he would head back to his law office, where new cases were already piled up. He would represent a transgender inmate brutally beaten by a cellmate, and later an African American father of twelve who was shot and killed by a University of Cincinnati police officer during a traffic stop. The university would pay millions to settle with the family, establish a memorial and launch police reforms.

But as the black convertible rolled past Fountain Square, where five same-sex couples had married the night before, Al decided that he would give himself the day to celebrate with his city.

Jim left Cincinnati after the parade and crisscrossed the country over the next several months. He went to the gay pride parade in San Francisco. He met with Edie Windsor in New York. He sat with Michelle Obama in Washington, D.C., during the president's final State of the Union address. He flew to Florida, where he looked out at the faces in a crowd from a group called Equality Florida and mused about his new life, how he had found purpose and passion in loss. "I'm still getting used to what happened on June 26 and my place in it. People call me a hero, a pioneer, a courageous man. I don't think of myself that way. I simply think of myself as someone who was lucky enough to fall in love and keep my promises to the man I loved. I'm also someone who finally found his voice."

He had no idea where his life was heading, but he knew he was drawn to new frontiers in the gay rights movement now that same-sex marriage was a constitutional right in all fifty states. He posted messages on Facebook and Twitter about the need for a federal anti-discrimination law protecting sexual orientation and gender identity, expanded support for transgender and nonconforming people, and lawsuits and lobbying against broad religious exemptions, much like the one raised by the county clerk in Kentucky who had refused to issue marriage licenses to same-sex couples even after the Supreme Court said she had to. Jim thought of himself as something of an accidental activist, a role he could never have imagined two years earlier, when he was a private man facing a brutal disease and the loss of a man he had loved for nearly twenty-one years.

Once, two months after the Supreme Court decision, Jim returned home to Cincinnati to help with a wedding. Kevin Cox had fallen in love with Tom Young two decades years earlier over long walks in a Cincinnati park that had once been home to a grand nineteenth-century estate. They wore rings but had never married because Ohio wouldn't let them, and a ceremony in any other state seemed pointless.

Cox, who has clipped gray hair and a solid build, had been home from his job as the lead designer at the city's weekly business journal on the day of the Supreme Court decision, and when Young came in late that afternoon after a day of nursing at a Cincinnati hospital, Young said, "Did you see the news?"

"Of course," Cox replied. "They're having a rally at Fountain Square and they're going to perform marriages. Unfortunately, we had to get a license ahead of time."

"Well," Young said, smiling, "what have you been doing all day?"

They had never met Jim Obergefell, but Cox sent him a message on Facebook two days later: Now that we can legally marry, we'd like to take the plunge. We'd be very honored if you would consider officiating for us.

Jim had received his minister's credentials a month before the Supreme Court ruling but had never performed a wedding. In late August, Cox, Young, Jim, and a handful of family and friends gathered under a pavilion with eight white columns and a domed roof, called the Temple of Love, in the same park where Cox and Young had spent hours walking with their dogs in the early 1990s. Cox wore a pink tie; Young wore purple. A cellist played the hymn "Simple Gifts." They brought their dogs, Grace and Sofie, and a wedding cake made of white chocolate.

Jim stood before them. “For many years to come, you will remember this day, yet beyond the flowers, beyond the music, beyond the expressions of joy and encouragement, may you celebrate through the years that which is most fundamental about your union. You are joining because you love each other, you respect one another and you will honor each other, and because you are committed to sharing equally in both the triumphs and trials of the days to come. As you begin this new chapter in your lives, remember: all of your yesterdays have led you to today, and your love will lead you into tomorrow.”

Jim looked at Cox. “Take this ring and place it on Tom’s finger.”

Jim looked at Young. “Take this ring and place it on Kevin’s finger.”

Cox had worn his ring for the better part of twenty-one years, taking it off for the first time only a few days before the ceremony. Young had trouble pushing it back on, and Jim threw his head back and laughed.

Then he said, “You have chosen each other, declared your love and purpose before family and friends, and have made your pledge of love to each other, symbolized by the giving and receiving of rings. By virtue of the authority vested in me by the State of Ohio, I pronounce you legally wed.”

ACKNOWLEDGMENTS

THIS BOOK would not have been possible without the support of Al Gerhardstein, who shared his time, memories, documents, and guidance. Al's wife, Mimi Gingold, and children Jessica Gingold and Adam and Ben Gerhardstein, helped tell Al's story and also the story of gay rights in Ohio. Thank you to the people of Gerhardstein & Branch for their patience and great support, especially Jennifer Branch, Sydney Greathouse, and Mary Armor.

The case for marriage equality didn't start and end in Ohio. We received tremendous support from the many co-plaintiffs and lawyers in Kentucky, Michigan, and Tennessee, especially attorney Regina Lambert, a passionate and gifted storyteller; Sophy Jesty and Val Tanco; Pam and Nicole Yorksmith; Michael De Leon and Greg Bourke; and Joe Vitale and Robert Talmas. Thank you for opening your homes, sharing your experiences, and introducing your children, including Grayden and Orion Yorksmith, Isaiah and Bella De Leon, and the dapper Cooper Talmas-Vitale.

Thank you, David Michener, for sharing memories of your late husband, Bill Ives.

Special thanks to those who provided legal and historical guidance, particularly James Esseks with the American Civil Liberties Union, who patiently offered detailed descriptions about laws, lawsuits, and court rulings. Thanks also to Susan Sommer at Lambda Legal as well as Freedom to Marry's Evan Wolfson and

Marc Solomon, whose 2014 book *Winning Marriage: The Inside Story of How Same-Sex Couples Took on the Politicians and Pundits—and Won* provided an extensive timeline on marriage equality in America.

Our sincere gratitude to the people of Cincinnati, particularly lawyer Scott Knox, journalist Dan Hurley, city council member Chris Seelbach, neurologist John Quinlan, minister Sharon Dittmar, and Kevin Osborne, the director of communications and LGBTQ liaison for Mayor John Cranley. Years of skillful reporting and photography by the *Cincinnati Enquirer* were essential to this book project, in particular the work of writer Julie Zimmerman and photojournalists and multimedia producers Carrie Cochran and Glenn Hartong, whose emotional video of the wedding of John Arthur and Jim Obergefell helped bring this story to the eyes of the world.

The exceptional work of other journalists also deserves mention, including stories, photos, and video that appeared in *Cincinnati Magazine*, the *New York Times*, the Huffington Post, Buzzfeed, *People*, SCOTUSblog, and the *Washington Post*. Thanks especially to the *Washington Post*'s Michael S. Rosenwald and Robert Barnes for ongoing coverage that provided critical perspective and detail.

Thanks to the ALS Association for guidance and information.

John Arthur and Jim Obergefell's many friends and family members were generous in sharing their memories, particularly Curtis Arthur, Keith Cassidy, Paulette and Mike Roberts, Ann Hippler, Meb Wolfe, and Kevin Babb. We are also grateful for the cooperation of the Ohio Attorney General's Office and for the invaluable insight offered by federal judges Tim Black and Martha Craig Daughtrey.

Our sincere thanks to the talented Jordan Rudner, for months of dogged research and reporting, and to Julie Tate, for bringing her immense research and fact-checking skills to this project under a tight deadline.

We would like to thank our agent Joelle Delbourgo for falling in love with this project even before the U.S. Supreme Court ruled in favor of marriage equality, and Jenny Meyer, for her passionate representation abroad.

Film producers Wyck Godfrey and Marty Bowen of Temple Hill Productions had a vision for a film based on this story, even before book rights were sold. Fox 2000's Elizabeth Gabler and Drew Reed provided crucial early support, for which we are deeply grateful.

Thank you to the smart, skilled, and supportive team at William Morrow for the passion and intense interest in seeing this story to print, particularly Liate Stehlik, Lynn Grady, Sharyn Rosenblum, Shelby Meizlik, Chloe Moffett, and, most of all, editor David Highfill, who loved this story from the start and provided wise counsel and a keen and sensitive ear at every stage of this project.

Jim owes Michael Volpatt a debt of gratitude for his friendship, support, and PR/marketing expertise over the past year.

Finally, we thank our family and friends for their support and faith: Jeffrey Rohrlick, Brett Cassidy, Zack Cassidy, Renee Cenziper, Michael Sallah, Jason Grotto, Lauren Levitus, Katharine Weymouth, colleagues at the *Washington Post*, Ann Hippler, Chuck Obergefell, Bob Obergefell, Bill Obergefell, Rich Obergefell, and families. Jim and John were fortunate to have support and love from many people, especially when John was ill, and throughout the fight for marriage equality. Thank you for being there, in good times and hard times. Your love helped to bring this book into being.

ABOUT THE AUTHORS

DEBBIE CENZIPER is a Pulitzer Prize–winning investigative journalist with the *Washington Post.* Over the past twenty years, she has investigated government fraud, public housing scandals, white-collar crime, and deaths in psychiatric hospitals.

Her stories have prompted congressional investigations, criminal convictions, new laws, and the delivery of hundreds of millions of dollars in funding for the poor.

She has won many major prizes in American print journalism, including the Robert F. Kennedy Award for Journalism and Harvard University's Goldsmith Prize for Investigative Reporting. She won the Pulitzer Prize at the *Miami Herald* for stories about affordable housing developers who were stealing from the poor. Debbie grew up in Philadelphia and graduated from the University of Florida in 1992.

She first met Jim Obergefell and John Arthur twenty years ago—Debbie's then-husband and John Arthur were first cousins. Debbie last saw John Arthur in 2011 at a family event.

JIM OBERGEFELL is a civil rights activist who embraced his newfound role after the U.S. Supreme Court made marriage equality the law of the land on June 26, 2015. A former Cincinnati realtor, he has worked with organizations such as the Human Rights Campaign and Equality, Ohio, and has been honored with awards from organizations such as Services and Advocacy for Gay, Lesbian, Bisexual and Transgender Elders (SAGE), Equality Florida, Equality North Carolina, the ACLU of Southern California, Cleveland Stonewall Democrats, the International

Court Council, the National Gay and Lesbian Chamber of Commerce, and the Ohio Democratic Party. *Foreign Policy* magazine named him one of its 2015 Global Thinkers, and *Out* magazine included him in its 2015 Out 100 list. The *Washington Post* called him "the face of the Supreme Court gay marriage case," and NPR referred to him as "the name that will go down in history." He earned a degree in Secondary Education and German from the University of Cincinnati in 1990.